Where the Blind Horse Sings

Where the Blind Horse Sings

Love and Healing at an Animal Sanctuary

Kathy Stevens

Skyhorse Publishing

All of the events and characters in this book are real. In a few cases, names have been changed at the subject's request.

Skyhorse Publishing books may be purchased in bulk at special discounts for sales promotion, corporate gifts, fund-raising, or educational purposes. Special editions can also be created to specifications. For details, contact the Special Sales Department, Skyhorse Publishing, 555 Eighth Avenue, Suite 903, New York, NY 10018 or info@skyhorsepublishing.com.

www.skyhorsepublishing.com

10 9 8 7 6 5 4 3 2 1

Paperback ISBN: 978-1-60239-669-2

Library of Congress Cataloging-in-Publication-Data
Stevens, Kathy, 1958-
Where the blind horse sings : love and healing at an animal sanctuary / Kathy Stevens.
p. cm
Includes bibliographical references
ISBN: 978-1-60239-055-3
Catskill Animal Sanctuary--Anecdotes.
1. Horses--New York (State)--Anecdotes. 2. Domestic animals--New York (State)--Anecdotes. 3. Animal rescue--New York (State)--Anecdotes.
SF301 .S84 2007
2007010143
636.08'3209747 dc22

Design by Laura Hazard Owen
Photos by Michael DiPleco, Dion Ogust, and Julie Barone
Printed in the United States of America

Dedicated with love to

Dad and Grannylou,

two irrepressible spirits

Contents

Acknowledgments

I thank my parents, my brother Ned, and my sister Ellen for the magic of my early years—years that led me right back to where I started from.

§

Catskill Animal Sanctuary exists because of Jesse Moore. Thank you, JM, for all those wonderful years, and for enduring cow poop and country for as long as you did.

§

CAS is a place of, by, and for the community, and nowhere is this more apparent than in the efforts of five tireless heroes—Julie Barone, April Harrison, Lorraine Roscino, Walt Batycki, and Alex Spaey—who bring the best of themselves to this challenging work each and every day. I love and appreciate them more than words can express. The same is true of Gretchen Primack, Jean Rhode, and Chris Seeholzer. Who needs a big board of directors when she has three people of such energy, talent, and passion? Thank you to our small army of volunteers—people who have shared time, talents, hearts, and muscles with us, and to the hundreds of community businesses that have shared their products or services so generously. Special thanks to Walter and Charlotte, Joyce, Dee and Paula, Kelly, Pat, Jane, Dick, J.C., Elena, Allen, Cameron, Eileen, Karen, Gary, Kory, Elaine, Betsy, Gus, Jan, Melanie, Frank, Carin, Kirsti, Chris, Cathy, Julie, Bob, and Michelle, and to Jenny on the Spot, the Mid-Hudson Vegetarian Society, Catskill Mountain Coffee, Candle Café, MooShoes, Teany, Jivamukti Yoga School, PDQ Printing, and Kiel Equipment. Thanks also to Steve Rucano of Secure Construction and Frank Tiano of Tiano's Excavation for many, many kindnesses. To Jill Spero: thanks for your extraordinary leap of faith, and for your generosity and patience through our infancy. And to my neighbors at the top of the hill: I love you both and am so glad that you can continue to be a part of this place that you and Charlie created. Finally, to those whose smiles, kind words, and gentle touches brighten the lives of animals: keep offering these gifts—they mean more than you know.

§

It has been nothing but a pleasure to work with everyone at Skyhorse Publishing, particularly Nick Lyons, Bill Wolfsthal, and the tireless and talented Laura Owen. My deepest gratitude to Skyhorse for giving me a chance to share this story.

§

In the midst of a wonderful but challenging life, David Cooper is my touchstone. Thank you, David, for your love, patience, and support, and for reminding me to breathe. Thank you, too, for sharing the adventure.

§

Finally, to my dear friend Rachel Jacoby: I hope you are as proud as I am grateful.

Kathy Stevens

Foreword

When most of us say we love "animals," we believe we are speaking of all animals. In reality, though, we are selective. We certainly love dogs and cats, and may have a special love of horses. Most of us would also include "wild animals." However, we unwittingly have favorites. If asked to name the animals we have in mind, most of us would start with mammals (but probably not rats and mice), and then might mention birds (but perhaps not crows, pigeons, or starlings). Some of us would put in a good word for reptiles, a few of us would include amphibians, and a small minority might name fish, especially tropical, freshwater fish. But except for bees and butterflies (as adults, not as larvae) almost none of us would include insects or other invertebrates. In short, we discriminate. Our love of animals turns out to be lavish affection for relatively few. As for the others, we seldom think about them.

Perhaps the most ironic aspect of our discrepancy applies to the birds and mammals that we eat and otherwise use—the farm animals. I spent part of my childhood on a farm and therefore once cultivated a compartmentalized attitude toward farm animals, feeling great affection for the farm dogs and cats, also for the dairy cows and the work horses, but little or no affection for the steers we sometimes raised for beef, or for the pigs, sheep, and chickens. We named the dogs and cats, the cows and the horses, but were forbidden to name the others. Instead, telling ourselves that they had no intelligence or emotions, we killed and ate them. My experience was far from unusual—it is very difficult indeed to eat a slice of ham cooked with honey mustard and decorated with a sprig of parsley, and at the same time also consider the personality of the individual that provided it. This is as true of our population in general as it is of farm families. If we eat meat or wear leather, we must compartmentalize. We must erase the knowledge that our food or leather garment is the muscle or skin of a sensitive, intelligent, emotional being. The shrink-wrapped meat that we buy in the supermarket cannot, we feel, have a former identity. Meanwhile, we continue to cherish our pets and donate money for wildlife conservation.

Thus Kathy Stevens is of particular interest, as she has seen what most of us seem not to see—the personae of farm animals. She created the Catskill Animal Sanctuary for them, one of very few such sanctuaries in the country. She cogently points out that the meat industry encourages our lack of concern—if we knew who it was we were eating, and what the animal had to suffer, we animal-lovers might feel protective. But most of us would rather not know, and thus we do little to help farm animals. So Kathy has set about to counter our apathy. Part of the mission of the Catskill Animal Sanctuary is to educate the public, hence tours are offered so that visitors can meet the animals they might otherwise ignore or be eating. And she has written this book to tell their stories. Given her sensitivity to animals, every story is a perfect gem, and all are intensely moving. I found one story of particular interest—the story of Rambo, a sheep so fierce he would "rear up on his hind legs and come full tilt" at whoever was unfortunate enough to have to clean his pen or feed him.

When I lived on the farm, sheep seemed to be mindless creatures. We used the expression "silly as a sheep." Back then, Rambo's ferocity would have supported my impression. But later, as a young adult, I met a flock of semi-wild sheep in Scotland. To my surprise, their intelligence and knowledge shone. This was especially true of the leaders, the older sheep, and of a wise old ewe in particular. Thus I realized that the sheep I had previously known were (a) young and inexperienced, (b) leaderless, and (c) captives in a controlled environment. They had no way to learn, and no one to teach them. They were understandably anxious and easily frightened. Of course they seemed mindless. After my eyes were opened to sheep, I looked for examples of their intelligence, and have found none better than the story of the fearsome Rambo who, after living in the Sanctuary, was calmed by Kathy's care and kindness to the point that he was able to understand his surroundings as well as she did. On his own, he purposefully saved the lives of some of the other inhabitants. I won't spoil the book by telling his astonishing story. I'll just say that it rings perfectly true. A person or an excellent dog might act as Rambo acted. I can also say, categorically, that if Rambo hadn't found Kathy, or vice versa, he would have been shot, and his life and amazing ability would have been wasted.

If only one of the stories herein had been told—especially the story of the blind horse of the title—the book would have achieved its goal of education. We cannot read about this horse without opening our hearts to him. Hence the cumulative effect of all the stories, and of the history of the Catskill Animal Sanctuary, is powerful indeed. Of the dozens of books about animals that are published every year, very few change the way we think about them. But after reading this book, we will think differently about farm animals. With their complex intelligence and empathetic behavior, they have much to teach us. We can ignore the identity of a serving of meat as long as we know nothing else about it, but we can't ignore the identity of our teachers.

Elizabeth Marshall Thomas
March 2007

Where the
Blind Horse Sings

Introduction

Nine of us are climbing a narrow path up a Pennsylvania mountainside. Granted, this is Pennsylvania (we're not talking Kilimanjaro), but the climb is still rocky and uneven, and our hands are full—we will likely be tired before the work even begins. Troy has a large dog crate balanced on his shoulder while Walt carries backpacks filled with rope, sheets, lead rope, and a first aid kit. Volunteer Sharon Ackerman, an acupuncturist, carries water for the troops and the rest of us pass heavy crates between us. I carry one atop my head for a while, then pass it on to volunteer Vanessa VanNoy, who's taking a day off from yoga instruction. Catskill Animal Sanctuary's animal care director Lorraine Roscino doesn't even work at the Sanctuary on Wednesdays but here she is, climbing with us on her day off, as is Friday volunteer Debbie.

We are going to rescue nine feral goats. At least one is injured. Four are pregnant. All are emaciated.

"I see one of the black ones," says Gray Dawson, who is just ahead of me. His wife Melissa has left the group and gone ahead of us. It was Gray and Melissa who called us about the animals. Melissa has seen the goats five days in a row now at the top of her friend's large mountainside property, and per our instructions, has been taking grain to them, hoping to earn their trust.

The rest of us hang back in a pod, sitting very still, and sure enough, two pygmies come bounding down the mountain. Walt moves forward. In an instant, the two girls have lead ropes around their necks and are munching happily on grain. We lure them into the biggest crate, knowing we'll need all available hands to safely catch the seven remaining goats.

"We'll never get them," Melissa says dejectedly. It's the pygmies who have come to her several days in a row. Of the seven remaining, only one has come cautiously within her reach in preceding days. For the others, even though their emaciation is extreme, fear has consistently won out over hunger.

I am far more confident than Melissa, and though we don't talk about it, I suspect the rest of the CAS crew is, too. It's not like we've had a lot of experience rescuing wild goats from a mountainside. But we have taken in hundreds of animals who were originally, though usually not for long, terrified of us. We know how frightened goats move. We know they can spring and leap with remarkable speed. Our bodies know how to match their movements.

The injured goat and her very pregnant friend are huddled at the edge of a vertical rock face. There's a thirty-foot drop to the craggy, boulder-filled forest floor. If a terrified goat leaps, she'll die; if a human slips, she'll die.

"Should I try to get behind them to drive them down the hill?" Melissa asks.

"Yes," I say, and as she does so, the rest of us very, *very* slowly encircle the two frightened animals.

A boulder juts upward between Lorraine and me—a boulder that a frightened goat will very likely use as a springboard to leap over our heads and out to freedom. "Kathy," Lorraine says. I turn; she's found a long, large branch. We stretch to hold it between us, hoping an additional visual barrier will dissuade a fleeing goat. "Great!" I thank her.

We close in. We are all crouching; we are all quiet; we all utter soft words of encouragement. We are fifteen feet from the goats. Then we are twelve; then we are six. A few more feet and we can dive and catch them. But a goat flies through the gap between Lorraine and me. I leap for her. We collide in a heap to the ground and are sliding downhill toward the cliff. Instantly Walt and Troy are there, wrapping the frightened animal in ropes and sheet to keep her both safe and contained. I sit up.

"Good tackle!" a voice from the left says.

"Thanks," I say, quickly dusting myself off.

Over the next hour, we continue this methodical work until just one goat remains. She's below us, backed into a corner below the rock face.

Troy scales down a steep slope, followed by Debbie. I can't see the goat.

"If all of us come down, can we trap her?" I ask.

"I think so," Troy responds.

One by one, we lower ourselves into the small cavern. This time, the goat can only move forward; the rock wall behind her is sheer and solid and completely vertical. We form a line and inch forward, our hands outstretched to close in the gaps. The goat, rail thin but pregnant, turns quickly in every direction, desperate to find a way through. She leaps but Troy leaps faster, they fall together and then Walt is there, safely harnessing her and wrapping her terrified, weakened body in a sheet.

It's clear we'll need to carry her out . . . so we do.

Three hours after we arrive, nine goats are safely loaded into a rented cargo van and are heading north to Catskill Animal Sanctuary. They will live in a large, hilly pasture filled with trees and boulders much like the forest they came from. Their babies will not fall prey to hungry fox or coyote. In this new home, they will have plenty to eat, warm shelter in the winter, and if they choose it, plenty of love from always willing humans.

Much has changed since *Where the Blind Horse Sings* was first published in 2007. Devoted supporters from around the world have enabled us to expand our infrastructure and build many new barns and shelters so that we can continue to say, "You have a home here" to farm animals who've run out of options. Several foundations provided the funding for our 100% conversion to solar power, so that CAS may well be the first sanctuary in the country that is completely off the grid. Partnerships with educators and progressive businesses are helping us expand our educational programming and bring greater numbers of children to Catskill Animal Sanctuary to experience farm animals as they've never known them: personable, inquisitive, audacious beings as individual as we are. By the time my second book is published in 2010, our organic garden will be feeding staff, volunteers, animals, and visitors.

Of course, not all the changes have been happy ones. For animal lovers, the brevity of our animal friends' lives is one of the toughest parts of our own lives. Over and over, we love and say goodbye, and it never gets easy. This is certainly the case at Catskill Animal Sanctuary, where we've lost many beloved animals, including several featured prominently in this book.

But as much as CAS has grown and changed, much remains almost poignantly the same. Our staff, for instance. Lorraine Roscino, who joined us as a timid volunteer terrified of horses the first year we opened, is now our animal care director and a crackerjack horse trainer. April Harrison, Julie Barone, and Walt Batycki arrived in early 2003, all as barn volunteers. During their first six months, both Julie and April were virtually too shy to speak. Julie is now my right hand and my left brain. As we grow, she oversees a host of complex tasks, multitasking with an efficiency that's a marvel to witness. April, meanwhile, runs the barn, directing each day's crew with confidence and focus, her eye always on the prize. Walt Batycki has morphed into our outreach guru, managing volunteer outreach, educational programming, events planning, and all our marketing efforts. Alex Spaey arrived in 2004. He is our workhorse, completing any task that requires strength and/or knowledge of building and heavy farm equipment. Whether it's nine degrees or ninety, Alex never quits.

I love these people. They are the heart of Catskill Animal Sanctuary; they are love in action. That they are all still here says something wonderful, I hope, about all of us. Sure, we often butt up against each other as we grow. But *we do grow,* together, and together we nurture this place we all cherish.

The most dramatic "sameness," though, is in the lack of sameness in rescue work. As Troy Gangle, our newest staffer, commented the other day, "You can't make this shit up." Troy was referring not to the fact that six of us left CAS to trudge up a mountain to attempt to rescue nine goats in dire straits. No. That was today. He was referring to *yesterday,* when a different group of us had to move four cows, one of them blind, from a pasture by my house to one a good half-mile away. Let's just say that while no one was injured, the move didn't exactly go according to plan, and I was breathless with laughter as six "expert" animal handlers alternately led, pushed, pulled, chased, cajoled, and lured the four cows—and lifted one large cow butt out of my personal flower garden. (That was the butt of Helen, the blind cow, whose response to being moved was, well, to sit down right where she was, which happened to be in my flowers.) It's true: If I've learned anything in eight years of rescue work, it's that there's no such thing as a predictable day. Those who need routine need not apply. For the rest of us, there's joy and richness, love and hilarity in working with resident animals who know in their bones that Catskill Animal Sanctuary truly is their safe space, their haven, their home.

Each day on this hallowed ground is an opportunity to witness miracles. One morning an elderly sheep nurtures a turkey. On another, we humans giggle as Murphy the dog and Ozzie the pig play tug-of-war with a blanket in the middle of the barn aisle. On a summer afternoon, I whisper a wrenching goodbye to my dear friend Paulie, who, as he's dying in my arms, rubs his tiny rooster head on my forearm, over and over again. Paulie couldn't hug me; he had no arms . . . but how easy it was for him to say "I love you" in his final moments. I now unequivocally know something I didn't know in 2001, the year I turned down the principalship of a new school in order to begin this new adventure. Whether feathered or furred, two-legged or four, human or horse, we're all the same. The animals announce this truth in myriad ways many times each day, and it's a lesson that has forever and irrevocably informed all that we do here.

As the world teeters on the brink of calamity, there is this little slice of heaven called Catskill Animal Sanctuary, and I wonder, indeed, what I've done to deserve it.

Where the Blind Horse Sings

Part One
First Steps

- 1 -

A Diamond in the Rough (One)

A year after we moved into our temporary residence, we began in earnest the search for a permanent home. Our criteria were simple: seventy-five acres or more, a few pre-existing shelters, and a house on the premises. For our own convenience and sanity (and that of future volunteers), we also hoped to be closer to either Kingston or New Paltz, two small cities that were both thirty minutes from our current home. The area where we lived was nearly twenty miles from the interstate. One small, foul-smelling grocery store sold junk food and rotting fruit, and the only restaurant was a diner: hardly an option for vegetarians.

Much to our alarm, a million dollars or more was the going rate for farms of that size. No matter our early success at fund-raising, that kind of price tag was way out of reach. We quickly learned, too, that much of the value at that level was in the home that came as part of the package: wonderful eighteenth-century stone houses with six-hundred-square-foot living rooms, or nineteenth-century, five-bedroom clapboards with deep farmers' porches. There were only two of us. We didn't need a four-thousand-square-foot home. We kept looking.

Jim Nimal and his wife became members of Catskill Animal Sanctuary not long after we opened. Like everyone else in the community, Jim knew we were looking for a place to hang our shingle. Even so, I didn't expect what I heard when he called one June morning.

"I investigate welfare fraud," he explained, "and one of my cases is in Saugerties, near a former racehorse farm."

I perked up. A training farm would probably have a huge barn . . . now, that would be a bonus.

"I've gotten friendly with the owner, a guy named Charlie. He's pretty desperate to sell."

"Oh, really?"

Jim wasn't sure of the details, but he knew Charlie was anxious. With the next words out of his mouth, I was in my car and on my way:

"Kathy," Jim said cautiously, "it's in rough shape."

We'd seen quite a few places that were a little rough around the edges, so the comment didn't faze me. In fact, I was heartened, because I hoped that an eager seller plus a farm in disrepair would equal a reasonable asking price.

I pulled to a stop at the bottom of a winding, downhill drive and blinked for a long, long time. A mound of used tires—scores of them—sat directly in front of me. Behind them, a rusted horse van, an ancient excavator, a school bus on its side, an old camper, and cars from a bygone era had been brought to Fortune Valley Farm to die. The hill sandwiched between the top and bottom of the downward-sloped, serpentine driveway was evidently a massive burn pile, judging by the assortment of rusted, charred metal on its face. Box springs, metal chairs, paint cans, air conditioners, defunct lawnmowers, appliances of every stripe, including several stoves and refrigerators, had been dumped over the years by Charlie and his friends.

"Yeah," Charlie would later explain in his gravelly voice. "No sense paying to dump it."

But what a setting for a farm: gently rolling hills and an expanse of flat meadow—perhaps fifty acres—were flanked by wooded cliffs. Over thirty weeping willows surrounded a pond of about an acre, which was a mere hundred feet from a huge barn.

"Mind if I walk the property?" I asked Charlie, who sat in a blue vinyl chair perched outside the end of the barn.

"Go ahead," he said between long drags on his Kool cigarette.

The land itself was wildly overgrown with burdock, nettle, thistle, and loosestrife. The grass reached to my waist, making it impossible to judge the terrain. Though the fencing was rotten and largely collapsed, enough was standing to reveal how Charlie and his brother Frank had fenced the farm when they began a Standardbred training business in 1970. (In fact, the brothers had hand-sawed hundreds of the locust posts for their property after suppliers could no longer provide what they needed.) Thirty years later, however, this land would require far more work than a vacant piece of land would: months of cleanup by a small army, some serious demolition, and the rebuilding of roads would have to take place before we could think about construction. The pond was choked with algae. I surveyed the desolate scene, and was willing to bet that the well was seriously contaminated.

I hadn't even entered the barn. Why wasn't I driving away from this place?

Twenty stalls, ten on each side, flanked the wide center aisle of the big barn. Two large rooms, both roughly 25' by 40', had been extended from the middle, so that from the air, the barn looked like a giant plus sign. Though it lacked a hay loft, the barn had all the other prerequisites: spacious stalls still lined with rubber matting, a hay storage room, and an area large enough to double as a kitchen and feed room. But it took some doing to imagine that horses had once inhabited this building. El Caminos on cinderblocks, their hoods wide open, lined the aisle. Engines and tool boxes, tires and big metal hulking things that only car mechanics could name, thirty-gallon drums and barrels with assorted metal dotted the walkway. In fact, every step down the 135-foot aisle was impeded: walk around a rusty car, crawl under a group of planks on the four-foot aisle walls of the stalls. And then there were the stalls themselves.

"That's Jack's stall," Charlie said, pointing to the one that was jammed floor to ceiling with everything from framed prints to boxes of tools to cans of nails. Old mail and older clothing were strewn about. Everyone who knew Charlie knew the size of his heart, one that extended to people who needed to dump or discard old goods or to store the bits and pieces of a lifetime. Why rent a storage place, after all, when there were plenty at Fortune Valley?

Saddles and bridles, moldy horse blankets, and buggies were tossed into some stalls. "The Latrine, Past and Present" was the theme in other stalls: twenty toilets of various shapes and shades were tossed haphazardly in them. ("You need a toilet?" Charlie asked me at one point. Evidently he had been the recipient of dozens of new toilets when a local home-supply store closed. I declined his offer.) It was a toss-up as to which was thicker: the layer of dust that blanketed everything in the barn or the cobwebs in every corner.

Charlie was still in his blue vinyl chair, pulling hard on another cigarette. "It's all he does all day long," Jim had said. As I climbed over cans and generators and air pumps to make my way down to him, something just beyond Charlie caught my attention. In the midst of the ashes and desolation, rising in defiance of everything around them, fat red tomatoes and gorgeous green peppers—dozens and dozens—hung heavy on the vine. It was the single spot on this sad ground that spoke of something good.

"Charlie," I said, placing my hand on his shoulder. "How many acres is the farm?"

With his massive, tawny hand, Charlie pinched his cigarette butt into lifelessness and placed it in his shirt pocket. "You can have whatever you want," he responded.

"What do you mean?"

"Whatever you want. $75,000 for the barn, $3,000 an acre."

No matter how much work we would face to bring fortune back to this valley, we had looked enough to know that Charlie's numbers were fair. Indeed, even without a house, they were generous.

"Those are great prices, Charlie, but I'm asking how many acres the farm is: Sixty? Eighty? One hundred?"

"Like I said, Kat," he began. (From this very first meeting, I was always "Kat" to Charlie.) "You can have whatever you want—the barn plus two acres, the barn plus twenty, the barn plus sixty."

The light was dawning. Charlie really needed to sell, and he was evidently willing to subdivide pieces out for cash in hand. Whether this was legal, we'd have to find out—how much acreage existed, we'd have to find out. Charlie had broken up an original property of several hundred acres years before—"always with handshake deals"—and was uncertain of the size of the remaining property. All I knew at this moment was that I wanted this place that desperately needed rescue. The location was ideal; there was, I hoped, ample land; we could rejuvenate the beleaguered pond . . . we could bring fortune back to Fortune Valley. It didn't have a house. But no matter. Somehow, we would make this work.

§

The land, it turns out, was in two different towns: a fifty-five-acre piece was in the town of Saugerties and a twenty-acre piece was in the town of Ulster. A surveyor marked the boundaries for us. Charlie and his brother also owned another adjoining piece of land of roughly thirty acres; Charlie mentioned that this piece might also be available.

We did the math. Despite the work to convert a forlorn farm into a comfortable working sanctuary (it would be an adventure!), despite the contaminated well (we could dig another one!), despite light sockets that exploded into flames when one turned them on, despite the lack of a house (we could live in a tent! we could build a cabin! we could buy a modular—or a trailer, even!), Fortune Valley Farm was a hell of a deal. If my dad had died, this would be the place to write, "I can see my father turning over in his grave." But my dad was very much alive and in my life, and would have been mortified at the thought of my living in a trailer. Yet what was instantly clear in this moment of decision was how far down the list of priorities my own dwelling had slipped. Any old roof over my head would do. What mattered was sufficient space for the animals. (I even fantasized about building a space above the barn, but Jesse quickly squelched that idea. Never far from his city sensibility, he had no interest in hearing animals rustling in the hay beneath him, or, god forbid, in breathing in barn smell as he drifted off to sleep.)

"What do you think?" I asked Jesse.

"This is your gig, babe," the good man said to me. Translation: you're crazy, but I already knew that. Jesse would help with his whole heart for one more year, until it became abundantly clear that cow poop and country really weren't for him. Before he followed his muse to New Orleans, we had nearly sixteen wonderful years that I will always treasure.

Raising the funds to purchase Fortune Valley would be no easy feat. I approached Rachel Jacoby, a CAS member who had said from the beginning, "Let me know when you find a place—I'd like to help." Rachel was, and is, a generous woman, so I knew that she had a substantial donation in mind. Perhaps she'd buy the barn: certainly a $75,000 contribution would jump-start a fundraising campaign.

We sat in Rachel's backyard, each with a glass of wine. Murphy swam for sticks that I tossed into her pond.

"I've found our farm," I said with a smile.

I began with the bleak details—collapsed roofs, contaminated wells, cars and refrigerators and tires tossed on a charred hillside—and waited for Rachel to escort me to my car.

Instead, she smiled and said, "This sounds like fun!" The world, I thought, needs more people like this.

And then I described the place I knew Fortune Valley Farm had once been: the pond dug by the Tiano brothers and the willows planted around its edge; the long stretch of valley bordered on all four sides by woods; the remains of a half-mile racetrack that could be the beginnings of a road through the property's center; the big old barn in the middle of it all, which, though filthy and forgotten, stood as a monument to what had surely once been a magnificent place created by two brothers who loved animals, the land, and each other.

I described Charlie and his little dog, Duker, who never left his side, and the cast of characters that kept them company: Charlie's friend Jack, his nephew Billy. Others, all men with time on their hands, seemed to stop by to watch over Charlie and his tomato plants. "He's a sweet man and he needs to sell his farm," I said.

When I described the unorthodox pricing scheme, Rachel still didn't flinch. Instead, as I was, she was excited.

"Do you know what a deal that is?" she asked.

"Well, keep in mind that there's no house, but yeah, it's a good deal."

"How many acres?"

"Uh . . . that's the weird part." It was tough to get clear information from Charlie. He and his brother Frank had indeed sold off much of the original property over the years. Maps had been drawn and redrawn; Charlie honestly had no idea how much land he owned. "But he's willing to let us buy whatever we want, and the land I saw is at least seventy-five acres."

"So we could buy the entire place for $300,000?"

I thought I heard Rachel say "we"; I know I heard what she said next: "Kathy, I really believe in what you're doing, and that's about what I have in my checking account. Let me sleep on it, but I think I'd like to buy it for you."

And she did.

- 2 -

A Diamond in the Rough (Two)

My friend Joyce is a raven-haired beauty whose passion, other than animals and her partner Bill, is cleaning. Nothing—not even a toaster or blender—is visible on her kitchen counter. "Too much clutter!" she explains. When asked how often she cleans her house, she replies, "Constantly." So when Charlie kindly agreed to let us begin moving animals prior to our closing date, I instantly thought of Joyce. Mention a cleaning project and Joyce gets glassy-eyed with fervor. Her mania was just what we needed.

We rented a huge Dumpster for discarding metal and it was filled within a week. "Bring it back!" we instructed the company.

"What on earth is this?" was a frequent question in those first clean-up weeks.

"Who cares? Throw it out!" instructed Joyce. The woman was in her element.

From September through December of 2002, we kept the junk dealers happy and the recycling center busy. Goodwill and

Salvation Army probably had to hire extra staff to manage the van-loads of overflowing boxes delivered to their back doors. Sweet Charlie and his friends, though not exactly speed demons, did their part: the auto cemetery in front of the barn slowly began evicting occupants. First, the tired old excavator disappeared, then the horse van. One morning, much to my delight, I walked in to discover the barn aisle free of vehicles. Now Joyce could really get moving! Though he moved in "Charlie gear," every day Charlie made at least one trip down the road with his tractor and wagon filled with stuff. Though I couldn't comprehend why he wanted any of it, Charlie actually bought a box car in which to store his transferred treasure. "I like to tinker, Kat," he explained.

Getting the place into move-in condition would be a tall order. Right off the bat, we needed a new six-thousand-square-foot barn roof, roads, a parking lot, a new well, and lumber and labor for five pastures and shelters. Oh yeah, and we needed a house. Jesse would cover that detail.

As they always have, our wonderful members rose to the occasion. Money came in. Volunteers when we had them, staff when we didn't, pulled hundreds of rotten fence posts from the ground. A foundation was poured and our new home came down the hill, in sections, on a flatbed truck. An excavating company brought big yellow earth-moving vehicles down the hill and raced around the clock to move earth, install pipe to divert water from flooded pastures, replace collapsed septic tanks, and load the remains of collapsed shelters into trucks and haul them away. Still very rough around the edges, the new Catskill Animal Sanctuary was nonetheless ready for our thirty-eight animals.

§

Gabriella was a small Standardbred mare whom Charlie had kept after his horse-training business ended. Not only was she named after Charlie's granddaughter; Gabriella was also the daughter of Fortune Moy, the Tiano brothers' first horse and the farm's superstar. Charlie couldn't bear to part with her. She was his link to happier days—to days when Fortune Valley Farm was a showplace where, according to Charlie's friends, "you could eat your dinner from the barn floor."

Short and stocky, Gabriella was built like a tank—she looked nothing like the few lean, long-backed Standardbred horses I had met. For years, "Gabbie" had been a free agent. She roamed the crumbling farm; indeed, she roamed the entire valley, but always came back to the barn for visits. Fed doughnuts by Charlie's pals, she was the fattest horse I'd ever seen. Because she had a painful hoof condition brought on by, among other things, an improper diet, Gabriella spent much of her time lying down. It would take some work to bring this girl back to health. Yes, that's right: Gabriella and a small flock of chickens were to be "gifts" from Charlie to Catskill Animal Sanctuary. What would he do with them, after all? He lived in the city!

The little mare was thrilled to have company! Each time the trailer pulled down the driveway, she would hurry to the barn to see who had arrived. She greeted our three cows, Samson, Molly, and Sammy, politely, and did the same with our herd of goats. Instead of returning to the spot against the cliff where she rested each day, Gabriella began to stay closer to the barn. Fortune Valley was coming back to life and she knew it.

And then, oh joy of joys, the horses arrived. We had prepared three pastures for them. The first was a seven-acre pasture with a creek running through it and a few wonderful shade trees in the southeast corner. To this we added a 30' by 12' shelter. Bucky, Ivy, Chance, Sergeant Pepper, and Bobo would inhabit this field. Casey and Junior, two geriatric ponies, would live in a smaller adjacent field with its own run-in, while ancient Dino and blind Buddy would graze in a third pasture during the day and be brought back to the barn at night.

Gabbie moved toward the barn like a steam engine when she saw the trailer. Horses! Horses! I led Bucky into his luxurious digs while volunteer Val led Ivy, Buck's sweetheart. Both spirited animals, the two trotted over every inch of the field, taking it all in. For a moment, Ivy would put her head down to graze, but when Bucky moved too far away, she'd race gazelle-like to catch up to him. Bucky entered the shelter; Ivy followed. Bucky raced out, stood in the middle of his new home, and, ears up and head high, neighed mightily, claiming his new turf.

Meanwhile, Gabbie, still a free agent, was having a conniption on the other side of the fence. Faster than I'd ever seen her move, she galloped the fence line, tossing her head wildly. It had been years since she'd seen another horse. Desperately, she whinnied to the two other animals. Bucky charged over; a fairly benign hello occurred. Ivy was next. Ever the lady, she greeted the stranger politely and with reserve. Rather than returning the greeting, Gabriella flattened her ears and gnashed at the gentle mare. "I'm boss lady around here!" she made clear. So, okay, this relationship, and evidently those with other mares, would take some work.

The lovebirds were followed by Chance and Bobo, and then

by Buddy and Dino. Each time, Gabriella appeared out of nowhere; each time, she called to her new neighbors; each time, all the requisite horse greetings—snorts, stamps, squeals—echoed through the valley, after which Gabriella would prance off, tossing her head wildly.

And then, finally, stiff, old Casey and stiffer, older Junior arrived. The pair had lived in a junkyard and were wormy, anemic, and infested with ticks when we took them in some months earlier. Now, on a relative scale, they were pictures of health. Junior, dark and slight, was well into his thirties and looked like a woolly mammoth. He hobbled down the ramp and followed Lorraine into his new pasture. He had no interest in exploring; instead, having spent most of his life on dirt terrain amidst rusted car bodies, he buried his head in the grass. Casey was next. He was a portly, 14-hand, good-natured paint, and he and Junior were inseparable. How happy we had been to rescue these two from their abysmal existence; how happy we were that they had bounced back and could join us in our new home.

We were nowhere near as happy, evidently, as Gabriella. From "somewhere out there," she appeared. But this time, whether she liked Casey in particular or just felt a sudden, overwhelming need to join the herd, the short, squat mare wasn't trotting. Painful hooves aside, Gabriella was racing toward the fence and she wasn't . . . slowing . . . down. One hundred feet away, now sixty, now thirty, one more stride and then the brown body lifted from the ground, hind legs catapulting her up and over the fence to join, as if she had always been a part of it, the Catskill Animal Sanctuary crew.

- 3 -

Angry Man

Rambo the sheep arrived full of testosterone and rage. While his terrified harem of fourteen ewes cowered in the corner when we entered their stall, Rambo evidently thought his job was to kill us. It was a job he seemed to relish. "You're in our space, so now I have to kill you" was what his behavior suggested. He'd rear up on his hind legs and come full tilt at whatever hapless human was assigned to feed the sheep, his head lowered just enough that the base of his massive horns would slam into the thighs of his unfortunate feeder. So violent and volatile was he that we began to compare the bruises on our legs. Eventually, a volunteer named Matt won first prize with a shoebox-sized black-and-blue that spread over both thighs and made walking difficult.

We tried fostering Rambo out to pre-approved homes. Twice he went out and twice he was returned. Carla, Mario, and Kristina Picayo, their lives already filled with their own animals and many of our sheep, called one day to explain that they were reluctantly cashing in their chips. The conversation went something like this:

> Carla: "Kathy, I'm so sorry, but we can't keep Rambo anymore."
> Kathy [voice heavy with dread]: "Oh dear . . . What's going on?"
> Carla: "He's trying to kill us."

The Picayos are devoted animal lovers and their concern was mostly for the safety of the lambs they were fostering. In any event, it was pointless to protest. The killer was coming back.

Richard and Linda Rydant generously took a turn. This time, Rambo's target was their fence. Within days, he had smashed it to smithereens. "We're sorry, but we're so afraid he's going to wind up on the road and cause an accident. He could easily be killed," Linda said, her voice flat with defeat.

Grateful for the respite but knowing it would be our last, we took the raging boy back, and for months continued to collect wounds and share war stories.

An eventual legal victory granted CAS permanent custody of Rambo and his friends. The ewes and their many offspring were adopted by their delighted foster families, while we rushed to the phone to schedule Rambo's neutering—a desperately-needed procedure that couldn't be performed until he was legally ours. We hoped that our testosterone-free boy would renounce violence, but we were sorely—pun intended—disappointed.

After over a year without progress, we were more than a little discouraged. Consulting sheep breeders exacerbated the situation, since the consensus was that "Jacob" rams were "highly emotional" and "extremely dangerous." One breeder was alarmed that we had Rambo, and advised us to "put him down immediately" before someone got hurt. While I decided not to tell her about the bruises, I struggled with whether, for our own safety, it was indeed time to give up on Rambo. In two years, we had experienced dozens of transformations: broken bodies and broken spirits made whole again. This would be our first failure.

Enter Carol Smith, a tall, lean brunette with contagious warmth and quiet, disarming candor. One of our first volunteers, Carol still leads weekend tours, and solicits more CAS memberships than the rest of us combined. Everyone I know loves Carol, and while her earthy good looks don't hurt, it's her heart that people respond to. My ex once said about her, "I think she's an angel." He meant a real one, and it's a description that's fitting. Carol is also one of the wisest people I have ever known. So when she strode into the barn one morning and announced quite simply that Rambo "just needed more time," we knew we had to pay attention, bruises be damned.

And she was right. Over the next few months, the frequency and intensity of Rambo's outbursts decreased. His great eyes, the color of sun-touched wheat, softened. Did he finally know in his bones that he would never again be confined? Or did he, now that his friends had all been adopted, choose us—Walt, April, Lorraine, Alex, Julie, and myself—as his family?

Was it something else altogether? Our prisons are filled with people whose bleak circumstances lead them to violence. It's the rare individual who frees himself from his past and lives the remainder of his life with purpose and compassion. Like these rare individuals, Rambo was finally able to leave his past—prolonged starvation and years of imprisonment in a filthy stall crammed with animals—behind. It had taken nearly two long years, but Rambo was no longer a threat. Still, who he would become we could not have fathomed . . . not in our wildest dreams.

- 4 -

One Cold Bitter Night

Rambo's metamorphosis was as subtle as winter becoming spring. First came a letting-go of all the behaviors that protected him and his herd, which I believe were also expressions of anger about the conditions in which he'd been forced to live. Though Rambo quickly learned that at CAS he could count on tasty meals twice a day, spacious pasture and shelter, and kindness from his new humans, his responses were conditioned by years of deprivation. And then one day they were gone. To our utter delight, Rambo let go of anger and fully embraced the ethos of CAS. I first saw this on a cold November night.

Before we built our shelter to accommodate turkeys, our two big birds—blind Cliff and his protector Chuck—were taken out each morning to a spacious enclosure shaded by willow trees. There, the two pals would enjoy much that turkeys in the wild enjoy: grass and trees, bugs and grubs, the chance to stretch their legs in the warm sun, to preen lazily and relax with each other. We'd bring them back to their stall in the big barn at the end of each day. One bitter, wet evening, Murphy and I rushed to the barn for our nightly animal check. Rambo lay in the barn aisle just outside Dino's stall; we had long abandoned our effort to enclose him in his own space. (Rambo would literally ram the heavy door that kept him in until it either fell down altogether or enough of the heavy 2' by 6' boards it was made of loosened sufficiently for him to shove them out of the way and escape. He had been confined for far too much of his life. He simply would not tolerate it any longer.)

On this night, Rambo stood and looked me straight in the eye, wanting a treat. I obliged with a pear procured from the kitchen, then walked over to check on Chester.

"Hey, old man," I whispered to the ancient Appaloosa, his head low and eyes heavy with sleep. "Hi, goats!" I called softly to Amy, Noodles, VanGoat, PeeWee, and the rest of the goat gang, who rushed to their stall wall, jumped up, and in unison rested their hooves on the sill and reached out for what they knew awaited them: a bit of carrot, a pear, a plum, or an apple. Farther down the aisle, mighty Policeman, a thousand-pound pig whose name and size belied his gentle nature, snored happily in his heated stall. April or Lorraine had covered him and his pals Belle, Bart, and Napoleon with mounds of soft hay; only Policeman's snout was visible as the mound moved up and down with each breath. Bobo, an old blind mare, munched contentedly on her hay, but as soon as I called, "Bobo, right here, Bobo, I'm right here," she walked over for what she loved nearly as much as hay: chin scratches. The chickens perched side by side in their heated roosts, and Rosebud the cow lay as contentedly as only cows can in her quarantine stall at the end of the barn. On this night as on nearly every other night, all was well at CAS. Or so I thought.

As I did each evening, I turned around at the end of the long aisle and called good night to my friends. Like my singing to the cows on many nights, it was a ritual that I didn't dare perform in front of any human. Mimicking the Waltons, I actually called out the name of every single barn resident: "Good night, Belle! Good night, Claude! Good night, Sammy! Good night, Hampton!" and so on.

Before the first name left my lips, Rambo charged up to me, stopped dead, looked up with his great yellow eyes, and bleated. "Something's wrong," he said—no matter that what he said was "baaahh," because his communication was perfectly and instantly clear. It was, in fact, the first time I'd heard his voice.

"What?" I asked him. "Show me what's wrong."

The great proud beast marched halfway down the aisle and made a sudden ninety-degree turn into the empty turkey stall.

Oh my god, the turkeys! In the frenzy of a hectic day, we'd forgotten to bring the turkeys in.

Stunned by what I'd witnessed, but concerned about the birds, I'd have to absorb the import of this moment later.

I thanked our guard sheep effusively and ran out to the turkey yard, accompanied by Rambo and Murphy. Chuck cowered at one end of the long enclosure, his feathers drenched, his head tucked in a futile effort to stay warm. Poor Cliff, meanwhile, was outside their pen, motionless in the driveway in a cold, shallow puddle.

"Oh, my poor boy, I am so sorry," I whispered as I approached Cliff. Scooping the animal into my arms, I carried him into his dry, safe space, Murphy and Rambo at my side the entire time. The three of us repeated the process with Chuck. I toweled off the birds, kissed them on their rubbery heads, checked their food and water, and closed the door behind me.

It was time to thank Rambo. In the darkened, hushed barn aisle, I sat on my knees and looked deeply into his eyes. "Thank you, Rambo. Thank you for telling me about the turkeys. What a good job you did . . . what a good, good job." I took his face in my hands and massaged his woolly cheeks.

What had just happened? So much was revealed in that single communication. That a sheep was aware that the turkeys were outside was impressive enough. That he figured out how to tell a human blew me away. But most astoundingly, Rambo had just shown concern for two animals of a different species and had known that I would help them. I took his face in my hands—he allowed this—as tears began to fall. "Okay, boy," I said, looking into those eyes of his. "If this is who you are, you've got a big job ahead of you."

I received my graduate degree from Tufts University in 1989. In the three years of the program, I read over a hundred books by noted public-policy experts, politicians, historians, sociologists, teachers, and philosophers. The influence of a few of them—John Dewey, Noam Chomsky, Jonathan Kozol—on my thinking about education was profound. But somehow, the lesson I'd just received from a sheep far surpassed in its impact anything I'd read, discussed, or debated at one of the country's top universities. As brilliant and instructive as these thinkers' insights were, nothing they wrote challenged what I believed about teaching and learning or about the role of education in the lives of children. Nothing I read told me that my core beliefs were based on a false set of assumptions, on naïveté or ignorance.

But in a darkened barn on a bitter early winter night, a sheep who finally believed he belonged with us did exactly that.

-5 -

Boys, I Need Your Help!

In January of 2004, Pat Valusek of the New York State Humane Association called and uttered the familiar words: "We've got a situation. Do you have room?"

Two days later, forty animals were removed from one of the most desperate and deplorable situations I've ever seen. Their owner, an elderly woman, had numerous prior arrests for animal cruelty dating back to the 1970s. In 1994, 148 animals were confiscated from her property. Like most hoarders, she was at it again, feverishly collecting animals despite her utter inability to care for them. On her farm two hours north of us, where she lived in filth in a condemned house without heat, electricity, windows, or doors, she once again had scores of animals: horses, donkeys, mules, sheep, geese, llamas, dogs, and cats. A few days before the seizure, investigators had found a decomposing dog carcass hanging by the collar from the ceiling. Farm animal carcasses dotted the property. Apparently suspecting another arrest, the woman had burned a barrel full of dead cats the day before we arrived. When I rounded the corner of one of the barns, two dead llamas lay on the ground, their throats torn open by something stronger, smarter, or hungrier than they.

Fourteen hours after our day began, we caravaned back to CAS with eight miniature donkeys, ten miniature horses, a mule, two full-sized horses, and three llamas. While we knew absolutely nothing about llamas, we knew we could quickly place them in loving homes. Given the circumstances, leaving them behind was not an option.

Early the next morning, I walked to the barn to check the new arrivals. Much to my dismay, the llamas were . . . missing. The previous night, our large hay storage room had been emptied and lushly bedded for the three boys, who had shown their gratitude by smashing down the massive door and going for a stroll. They were nowhere in sight. It was just after 6:00 AM; no one would be here until 7:00.

More than likely, the llamas were in a large open field that bordered the southern end of our property. But what did I know about catching llamas, leading llamas, herding llamas? Nothing. Moreover, these animals, like the others with whom they'd arrived, were skittish. They'd probably see me and bolt in the opposite direction. Yet I had no choice, for the longer I waited, the farther they could roam from the barn.

Enter the great dog Murphy. My intrepid yellow Labrador retriever pursues all things canine with such passion and abandon that I still, after ten years with him, delight in his joie de vivre. His official title, in fact, is Director of Canine Pursuits. As a Lab, one of his favorite pursuits is to be at his human's side (i.e., completely underfoot) no matter what she's doing. And this particular morning, she was herding llamas.

"Come on, Murphy, I need your help!" I exclaimed, my voice oozing drama. From years of experience, Murphy knows that this sentence involves locating some sort of animal, and it's a job at which he excels.

We took off across the top of the large horse pasture. Murphy raced alongside me, periodically looking up as if asking, "Where are we going, Mom?! I don't see any animals!" The plan was to move undetected a good distance past the llamas, then to double back behind them. Hopefully, they'd be startled enough to form a nice single-file line and stroll cooperatively back to the barn. But again, I knew as much about llamas as I knew about loosestrife, the invasive flower that was rapidly choking large areas of pasture near the pond: zilch. Would the llamas panic and scatter like deer? Didn't know. Would they stand their ground like donkeys, staring at us as if we'd lost our minds? Couldn't say. Perhaps they'd charge like elephants! My mind played out various scenarios alternately hilarious, alarming, and worthy of a spot on *Wild Kingdom* or *America's Funniest Home Videos*.

We continued for several hundred yards, until behind us we heard . . . oh my . . . just what was it that we heard? Fighting for breath, I stopped and turned around. Over the top of the hill Rambo bounded, and I witnessed what I'd previously only seen in cartoons as a young child—he was leaping like a kangaroo! So it was true that the sheep's fastest gait is a wonderful catapulting of himself—a boinging through space. For the sheep in a hurry, all four legs land simultaneously, all four push off, propelling the ball of wool atop toothpicks skyward. What a sight! Rambo covered the distance in no time and stopped dead in front of me, and I realized that his evolution was continuing. Either hearing or seeing Murphy and me running across the field, Rambo, the instinct to protect woven into his DNA, knew something was wrong, and was rushing to our aid. Knowing once again that the processing of what I'd just witnessed would have to wait, I knelt down, thanked him for coming to help, and uttered with determination, "Come on, boys, we've got to get the llamas!"

The three of us tiptoed along the crest of the hill until we were a good fifty feet behind the llamas. Murphy was nearly convulsing with excitement! Although a vet once explained that a dog's shiver means either pain or fear, Murphy's shiver, which ranges from a slight quivering of his back end to a full-out near convulsion of every body part including eyelashes, happens in situations in which a bark would blow his cover. If he can bark, he doesn't shake, but when he knows that releasing his pent-up excitement would ruin his fun, then he's silent, bless his heart, but looks like a Parkinson's victim on some serious speed.

The moment of truth had arrived. There we were, the three of us, crouched in the high grass of a thirty-acre field: "Watch me, Murphy," I commanded.

"O-o-o-o-k-k-ay," his head vibrated.

I was about to instruct "Animals to Murphy's yard," since "animal" to Murphy means any animal other than deer, squirrel, or "big bird" (turkey) and "Murphy's yard" means the farm, or, in other words, "Keep your fanny on the farm and don't go raid the neighbor's garbage."

It was too late.

Murphy bolted to the lower edge of the field and circled behind the heels of a big, dark llama. He bolted, yes—but was I really seeing this? Rambo stood in front of the woolly llama, blocking his escape with a charge and a lowering of the head clearly understood by the frightened animals. The yellow dog and the great horned sheep worked with skill and knowing, forcing the three llamas to huddle, and then, once they were a compact unit, driving that unit toward the long narrow lane that led back to the barn. If a llama balked, either Murphy nipped at its heels or Rambo threatened with a lowering of his magnificent head, while I simply sauntered behind the group, marveling at the gift I was receiving.

The two animals made short order of what I thought would be a Herculean, if not impossible, task, and when we arrived back at the barn, the staff was pulling in for the official 7:00 AM start of our day. When the long lane ended at the open barnyard, a speechless Walt joined our effort, the llamas were effortlessly scooted back into the hay room, and two forever-changed human beings went about their day.

-6 -

St. Francis of CAS

Nothing we humans did could convince Charlie the pig that there was a big wonderful world worth exploring just outside his stall door.

Charlie was a surprise rescue. A woman called to say she needed to place her lovely Angora sheep because she was moving. When we arrived, not only had she already moved, she'd also failed to mention a dozen or so chickens and one morbidly obese pig locked in a tiny space and barely able to move.

"She never liked the pig," a neighboring caretaker explained to us.

Now, I never particularly cared for an unusually schizophrenic cat that we adopted as an adult from a local shelter. The name that she came with, Trouble, was dead accurate, if rather unfortunate. Trouble stalked, bit, scratched, and attacked Murphy, and downright tormented our other cats. She was not much fonder of the two of us. But would I have left her behind if we'd moved? I couldn't imagine leaving a plant behind, much less a highly aware, sensitive, and intelligent animal dependent on humans for his very survival. But there Charlie was, and so he, too, came to call CAS his home.

It seemed that all Charlie wanted was his own stall. Double the size that he should have been, Charlie had trouble standing, much less walking, and when he attempted to walk he dragged his massive belly with him like a punctured medicine ball. Probably from the obesity, his malformed front legs were shaped like boomerangs. We put him on a low-fat, low-calorie diet of mostly fruits and vegetables, and over the next eight months, the weight slowly came off.

Surely the weight loss made Charlie feel better physically. And what about the fresh bedding each day? What about the encouragement he received from humans: the belly rubs, the nudges out into the warmth of a spring day, the gentle hands, encouraging words, and patience that had worked for so many before him?

Nothing we did could convince this pig that life outside his stall door held so much promise! Day after day, week after week, month after month, Charlie lay in his own urine and feces, only moving outside when we forced him to in order to clean his stall. This diminished life inside four walls was the only life Charlie had ever known, and after months without improvement, we struggled with the hard questions. Did Charlie want to live? His choice—as a now fully mobile and fairly slim pig—to lie in his own excrement day after day indicated, to us, depression . . . or even despair, for despite their reputation as dirty animals, pigs are actually fastidious, and never soil their sleeping area if given other options. (In fact, though they have very generous sleeping quarters, our pigs don't use the bathroom at all during the night. Instead, the instant they're let out, they lumber to their respective, self-selected latrines to do their business. As for their penchant for mud, pigs have no sweat glands and very little hair, so lying in mud is their ingenious way of staying cool and preventing sunburn.)

In the barn, the difficult discussions began: was euthanizing Charlie a kinder choice than subjecting him to this life? He had no joy, no pleasure except eating, which for a pig he did half-heartedly. Charlie had no drive to do anything that other pigs so clearly enjoyed. Those of us who knew him best thought that Charlie considered life a burden. Maybe it was our job to release him from that burden.

The discussions took place over many days, as all the while we willed the little pig either the desire to live or the ability to make it absolutely clear to us who would order his death that this was what he wanted.

A calm and knowing sadness was what I felt one afternoon when Lorraine joined Rambo and me in Charlie's doorway. "It's time to let him go, Lorraine," I whispered. "We've done everything we can for this boy, and look at him. This is not a life."

Lorraine wept openly. "I know," she agreed. "I wasn't ready for a long time. I kept thinking there was something more we could do. But you might be right."

We stood, two women who believe so strongly in the power of love, with our arms around each other's waists, sad that we might be losing a friend, devastated that we'd failed him. Rambo stood with us.

"He knows," Lorraine said, looking down at the golden-eyed boy and scratching him between his horns before returning to the feed room to prepare dinner.

"Kathy," Lorraine hesitated. "What if we put him on glucosamine?"

I thought for a moment. Several of our older animals were on this joint supplement, but Charlie was a young pig who was no longer overweight. I didn't think it would work. But it was Lorraine—a woman who took a bullfrog to the vet for surgery on a pierced eardrum—who was asking, and I lacked the heart to say no.

I knelt down in front of Rambo and looked into his knowing eyes. "I'm sorry, Rambo, I'm so sorry," I said, then kissed his forehead and walked out.

Sometime later, my phone rang. It was Lorraine.

"Come down here right now! You won't believe it!"

I raced to the barn and approached Lorraine, who was standing in front of Charlie's stall. This time, though, she was smiling widely as she motioned to me not to speak. For there in the hay that we had fluffed for a pig who wanted to die lay a sheep who wanted him to live. Rambo, who so hated confinement that we never stalled him, lay in Charlie's stall just inches from the sad pig. Surely Charlie could feel his breath. My knees weakening, I thanked heaven, earth, stars, gods, and anything else listening for blessing us with this creature.

"Do you think he can help?" Lorraine asked.

"Yes. Yes, I do."

The next morning, the next afternoon, and the next night, the sheep who under normal circumstances would smash down the door of any stall that contained him barely left Charlie's side. He could have if he'd wanted to—Charlie's door stayed open to encourage him to come out and explore the world—but Rambo had a job to do, and was evidently going to see it through.

§

Day Four of Rambo's vigil. Too soon for the glucosamine to have kicked in. My phone rings. "Come down! Come down! You won't believe it! Run!"

I dashed to the barn, tears already spilling out of my eyes. Halfway down the aisle, hugging the wall for security, Charlie the pig had just stepped out into the world. Rambo stood some distance away.

Charlie was grumbling like hell, warning us to stay away. "Okay, all you assholes, this is what you wanted, but I'm doing it my way," he announced with every step.

From that day forward, Charlie has lived life on his terms, accepting nothing more from his human caretakers than breakfast, dinner, treats, and an occasional belly rub. Oh, how he growls if one tries to pat him. He shuns virtually all affection and dares you to invade his personal space.

But, dislike of humans aside, Charlie is pig through and through. Some mornings you'll see him sunning against an outside wall, other mornings at the hay room entrance. On sultry summer days he's at the pond's edge, or in the muddy waddle made by the big pigs. Sometimes he chooses to join thousand-pound farm pigs Belle and Policeman on the shady side of the shavings pile. When Charlie comes in for the evening, we're likely to find him either sleeping with the big guys or surreptitiously stealing hay from their stall to pad his already generous bed.

Occasionally, we have to search for him, because somehow a sheep who underwent his own miraculous metamorphosis convinced Charlie of what humans could not: that there really was a big, happy world waiting just outside his stall door.

- 7 -

All in the Family

Cutting through the hay room en route to the kitchen, I heard April swearing like a sailor. Now if it had been Alex, I wouldn't have been concerned. Alex loves to swear and has mastered the art. April, on the other hand, rarely utters anything worse than "darn." So when I heard words unfit for church erupting from her, my walk became a mad dash to the kitchen, where Rambo, Hannah, and Franklin—two sheep and a pesky piglet—were making a very hasty retreat. Franklin the piglet was grumbling his head off.

"What's going on?" I asked.

"The f***ing feed delivery guy didn't latch the f***ing gate!" she moaned. Using Rambo's horns as their tools, the three marauders had ripped open four fifty-pound bags of feed; grain, pellets, and sunflower seeds spilled out everywhere. To make matters worse, they'd torn the bottom bags of a truckload of feed stacked on pallets in tall rows, so sixty other fifty-pound bags had to be moved before the mess could be remedied.

"Hey guys, no one's looking! Let's MOVE!" I can just hear

the threesome conspiring. How frustrated they must have been to be caught in the act just as they dove into their bounty.

Petri and Darwin, the ducks. Chickens Imelda, Nutmeg, Julia, Paulie, Sandy, Oreo, Blackie, and Spotty. Rambo and Hannah, the sheep. And then the pigs: potbellies Charlie, Priscilla, Hampton, Zoey, and Rosie, and four of our nine big pigs. Twenty-one animals roam the farm when we're working. The other animals have very spacious pastures and stalls or outdoor shelters to enjoy, but these twenty-one are free to roam from seven each morning until the staff closes up the farm each night.

Now if they actually roamed the entire farm, we'd have way bigger problems than an occasional raid on the feed room. But they don't. With the rare exception of a pig who ventures out into a cow or horse pasture to root, the animals stay close to the barn. Usually, in fact, they are underfoot, which is both heartwarming and maddening. Try moving the tractor forward with a sheep lying in the aisle in front of you, a duck napping underneath, and Paulie the rooster grooming Charlie the pig as Charlie leans against a stall wall.

Do the animals scurry away when Alex starts the tractor? Of course not. They don't budge. After all, this is the place where animals take full advantage of humans whenever they're able. (This is the primary expectation of our residents, a fact that Rambo conveys during orientation, held monthly at a time and location we humans have not yet ascertained.) The cleaning crew stops what it's doing, checks for small friends who could be squashed, and picks up or "shoos" anyone at risk. Then, finally, the tractor inches forward to align the manure spreader with the next set of dirty stalls.

CAS may not be the most efficient farm in the world, but it might be the happiest.

Did we have a thing or two to learn from Petunia, our first potbelly pig! Though many of the derogatory phrases comparing humans to pigs are inaccurate, "eat like a pig" is dead-on. Pigs love to eat. Combine that with intelligence and willfulness, and a free-roaming pig is a recipe for disaster. But Petunia loved people and wanted to be with us, so putting her in a pasture alone was unacceptable.

She taught us well and quickly all we needed to know. After Petunia trotted off with a volunteer's lunch bag, our lunches went to the fridge. After she broke into the feed room and went to town on a fifty-pound bag of feed, a pig-proof lock was put on the gate that separated that area from the barn. And after she trapped an electrician on a ladder when she was in heat, we learned to indulge her with extra belly rubs on her amorous days. Ah, well . . . all in a day's work.

These twenty-one animals have taught us lessons that I'd never have believed possible. Many others have, too, but we learn the most from those who share our day most intimately. In many ways, we humans and these animals are very much a family.

§

A little brown duck named Petri became our eighteenth free-roaming animal during the summer of 2006. An unwanted gift to a college student, Petri had spent her life indoors. She was small and utterly inexperienced with other ducks, so we feared that our outdoor ducks, mostly large Muscovies and Pekins, would torment her. With "Efficiency Be Damned" as our motto, we welcomed Petri into the Underfoot family.

Petri was delighted to be a barn duck! She had no idea what a pond was, and ran in terror from it. She wasn't even interested in the kiddie pool we filled to encourage her to feel more like a duck and less like, well, a person. Petri only wanted to be with us. During our morning cleaning of the twenty barn stalls, she would plop down in the middle of the aisle, watch Lorraine, Alex, April, and volunteers scoop poop, and participate in the conversation. As stalls were cleaned and the tractor and manure spreader were moved forward to the next group, Petri waddled forward, too, picked a new spot on the floor, and sat down to enjoy the show. She was an enchanting addition to the family.

Petri was followed by Darwin. As flashy a duck as Petri was plain, Darwin was jet black with iridescent green and purple highlights—a fine duck mutt. Originally rescued from a butcher, Darwin lived happily with a family for some time, until they moved and were forced to surrender him. Though he had enjoyed flapping around in a swimming pool with his human family, he looked at our one-acre pond and saw the raging Pacific, and when he eyed the huge Muscovies with their red masks and mottled feathers, I believe he saw Satan and his minions.

How happy the little guy was to be free, safe, and surrounded by friends! He latched onto us immediately, to the point that Julie had to carry him around on weekend tours so he wouldn't be stepped on. I heard her introducing Zoey to a dozen guests one Saturday afternoon. "This is Zoey, a super-friendly and gentle potbelly," she said, bending down to rub the sweet pig. "Whaaaa..." Julie's feathered assistant added from his seat in her arms.

Anyway, the two ducks' friendship didn't happen overnight. At first, Petri kept her distance, rebuffing Darwin's eager overtures. Squeaking her disapproval, she'd nip at his legs and send him away whenever he entered her personal space. Darwin, acting with patience and respect, waited, and their relationship developed quite naturally. The two ducks were strangers, then acquaintances, then pals; within weeks, they were inseparable. If Darwin was waiting outside the kitchen for dinner, so was Petri. If Petri was tearing at the grass outside, so was Darwin. We smiled at this new friendship.

But what about the swimming thing? Whoever coined the phrase "like a duck to water" must have known very different ducks than these. If we'd had a "Least Interested in the Pond" contest, these two—not the chickens, not even the rabbits—would have won, hands down.

Over the course of several months, however, we noticed the pair venturing toward the pond. "Did you see Darwin and Petri?" Walt whispered one morning. "They're nearly there!" I tiptoed out, and sure enough, there they were, four feet from the water's edge, motionless—simply staring. Probably their minds weren't quite comprehending the message their DNA was screaming: "You're ducks, damn it!"

But the next day, moments before their duck genes packed it in, Darwin took the plunge. Well, not a plunge, exactly—more like a little scoot into the water, a few perfunctory paddles of his webbed feet, and a mad dash back to the safety of his stall, screaming the entire way, "I did it! I did it! I'M A DUCK!" We praised him effusively.

What a pleasure it was to watch this pair approach the water in their own time, on their own terms, moving from terror to interest to unmitigated joy. Within days, their swims were long and triumphant. We'd watch as they swam slowly, exploring the banks, gliding to the middle of the pond, and swimming in broad circles or serpentines. They danced on the water, and there's no doubt that they were celebrating as they experienced a part of themselves they hadn't known existed. The other ducks took the water for granted; swimming was no different for them than walking is for us. But it was for Darwin and Petri. It certainly was for them.

Eight days after their initial foray into the water, the two friends were happily integrated with the Muscovies and Pekins.

§

In recognizing that the healing process is different for each animal, CAS allows each one to heal in his own time, on his own terms. We watch them closely to determine what they need in order to be truly happy. That's the richness here: as much as we can, we provide each animal what he or she needs not only for his or her physical health but for his or her emotional well-being. Fortunately for us, most ducks choose to be with other ducks, most goats with other goats. The ones that don't—or can't—join the barn family.

When we allowed two former house ducks to become part of the Underfoot family, we didn't know what the outcome would be. That they discovered their "duckness" and ultimately chose to live like ducks was fine with us, just as it would have been fine, though more inconvenient, if they'd wanted to be barn assistants for the rest of their days. Our job is to love and nurture without expectation, and when an animal who has only known humans discovers a preference for another life, then we, too, celebrate his choice. We know full well that it won't be long before another animal shows up who needs more than shelter, food, pasture, and love, and who will join, if only for a time, the Underfoot Family.

- 8 -
Welcome, Buddy

Buddy's family believed he wanted to die. He'd lost his vision, and they simply didn't know what to do. But Susan Wagner of Equine Advocates, an organization that had helped place hundreds of needy horses in loving homes, thought Buddy had the will to move past his blindness. "Kathy, he just needs a little help," she told me. Susan agreed to insure that Buddy's vaccinations were all current, that his medical records came with him, and that an experienced hauler would bring him safely to Catskill Animal Sanctuary.

So, early on a Saturday evening, the sweet Appaloosa arrived from Long Island. Down came the trailer door and out backed the spotted rear end of a frightened old boy. I took the lead rope from the hauler and thanked him for the long, safe journey.

"Hey, Bud," I whispered, standing just inches in front of the terrified animal. "I'm right here." Only once had I seen a horse shivering in terror, and that was when Ludwell, my beloved childhood pony, saw a cow for the first time and mistook it for a Tyrannosaurus Rex.

This was different. Buddy had been housed in a small, irregularly shaped barbed-wire pen. Every time he'd moved, a barb would gouge the flesh of his chest, his flank, his neck. Eventually, for safety's sake, he'd stopped moving. For the two weeks prior to coming to us, he'd just stood in one spot, head hung low, even refusing to eat.

So here he was, standing before me, the neediest animal I'd ever met. He was an average-sized horse with a long back and unusually pronounced withers. He was a milk chocolate-colored bay with a generous Appy blanket across his rump. The coarseness and sparseness of his coat suggested worms. Buddy's right eye was completely cloudy; his left had been removed some years earlier. He was thin, covered with scars, shaking violently, and terrified to move.

"It's okay, Bud. I'm right here. It's okay . . ." Over and over I whispered, letting him hear my voice, know my smell, feel my touch. That was enough for now. For ten minutes we stood nearly motionless. I scratched his chest, massaged his neck and ears, stroked his scarred face. All of this, he allowed.

Still, though, he wouldn't take a single step. One of several volunteers on hand to welcome Buddy handed me a small feed dish filled with grain. If this didn't motivate him to move the twenty yards toward his stall, we would be in for a long night.

I shook the dish in front of Buddy. Boy, did he perk up! Scooping out a nibble, I held my hand under his muzzle.

"Want some dinner, Bud?" I asked.

Evidently, he did! I backed up two steps, holding the lead rope in my left hand, shaking the feed dish in my right. Buddy inched forward, wary. He stretched out for a second bite. "Good boy, Buddy! What a brave boy you are! What a handsome, brave boy!"

And so the slow journey toward the safety of a cozy stall began. I would inch backward a few steps, he forward a few, growing more confident with each sequence.

It took forty-five minutes, but we'd reached the four-stall barn. Buddy's stall was just six feet in front of us. But before he could relax, we would first have to navigate the four-inch step up to the concrete walkway that ran the length of the building. I took the dish and placed it on the walkway, then patted the concrete walk several times soundly. "Up, up!" I explained to the horse, aware that this word would be useful for him to know. Shake the dish, smack the sidewalk. Buddy inched forward, the front of his right hoof colliding with the step. He stepped back—frightened, but not terrified.

"It's okay, Bud. You're doing a great job!" I praised. We can do this, I thought, knowing that the only goal was to begin building this broken animal's confidence and trust. Everything I did and said had to reassure, so I stood up and moved to his left side, gently taking his halter. Jill, my assistant at the time, moved in front of Buddy. She shook the grain and smacked the sidewalk. Together Buddy and I moved, no more than three painstaking inches at a time.

Directly in front of the walkway a second time, I used another word that it seemed a blind horse needed to know: "Stop!" I said firmly, and pulled back slightly on his halter. He stopped.

After allowing Buddy a few moments to process what he was taking in, I asked Jill to begin the sidewalk-smacking ritual again. She did.

"Up, up, Bud!" I encouraged him. This time, Buddy lifted his hoof high enough to place it down flat on the walkway. Praise, up, up, shake the grain. Jill moved backward, her shins now resting on the sill of the stall. Buddy and I moved forward. He was on the walkway. A few nervous breaths and moments later, the horse who, according to his owner, had wanted to die, was in his stall. It had taken over an hour to move sixty feet, but Buddy wasn't shaking anymore.

However small, the victory seemed momentous.

- 9 -

Dino Finds a Friend

For the first two weeks after Buddy's arrival, I spent hours each day with him. I sensed that he'd felt abandoned at his previous home. Clearly, he'd felt helpless. If we were to give this animal back his life, he first needed to know that we'd never allow him to be hurt, and would only take the next step out into the world when he indicated he was ready.

Three days after his arrival, Buddy was ready.

The barn that housed Buddy and assorted other creatures was a simple 24' by 48' structure. Four 12' by 12' stalls formed the front half of the barn, and each stall had a Dutch door—the top and bottom halves of the door operated independently. These made wonderful barn doors, as one could lock the bottom to keep animals inside but open the top so that animals could look out and enjoy the fresh air. Another wonderful feature of the barn was that Dutch doors were also built into the back wall of each stall, and opened into the back room: an open space that served as feed room and office. My desk was only six feet from Buddy's stall, and I kept both halves of the door open so that Buddy could feel that I was nearby. Only a rubber stall guard the diameter of a hose stretched across the open space. If Buddy walked too far, the stall guard would stop him.

Despite this caution (I didn't want Buddy walking into the office, bumping the corner of a desk, and getting scared), both halves of his front door, the door that opened to the outside world, remained open as long as I was in the office. After all, here was a horse who for weeks had been petrified to move. I figured that as his confidence grew, we should encourage him to explore his environment, so leaving the door open to enable him to step out into the world was a way of doing that. Based on his reactions to date, there was no risk that he'd go careening out and run headlong into a tree. He'd inch out, tiny step by tiny step.

Here we were, then, with a lot of work to do! But what did I know about building the confidence of a frightened blind horse? Buddy was a first. There was no methodology, no training manual, no workshop, no expert. As with all the broken spirits that have arrived at our barn door in need of help, it's always just us and the critters. Humans who observe, intuit, and respond with love, patience, and encouragement, and animals who perceive that we are not only here to help, but also that we'll do it on their schedule.

So Buddy was lavished with attention and praise. When I cleaned his stall, I talked or sang to him and touched him often as I moved about the space. Members of CAS came to visit, bringing carrots, apples, and his favorite treat of all: peppermints. We played "hide the peppermint," and no matter where it was (even in someone's shoe), Buddy could find it. Yet what fascinated me most was how he so wanted to know each and every newcomer, whom he'd greet, if the person allowed it, by gently smelling him or her from head to toe.

§

Three days after his arrival, Buddy wanted to know more than just those who came to meet him. He wanted to know about the place where he was living. I looked up from a phone call one morning and Buddy was standing several feet outside his stall in what we would soon call "Buddy pose": all four legs directly underneath his body, head high and turned to the left, ears forward, trying to hear what he could not see.

Inside, I was dancing! Buddy—blind, terrified, "I want to die" Buddy—was venturing out, on his own, just three days after his arrival.

"Buddy!" I called as I moved carefully through his stall and out to him. "What a good job! What a brave boy you are!" Over and over and over we would say these simple words to this horse each time Buddy would take the next step, whether it was a simple step out the trailer or outside his stall or into a stream for a splash. But more on that later. For the moment, what I knew was that this animal was a great gift, and that as much as I might teach him, his lessons to me would be far greater.

I also knew that he was looking for a peppermint. Yes, that's right. Buddy knew he'd done something deserving of a reward, and the clown was playing "find the peppermint," poking and prodding my pockets, my hands, my sleeves, my waistline. I dashed into the office and brought back three, which he munched with glee and then moved against me with his big old jughead, pushing it into my back and rubbing up and down. I think he was saying, "Okay, partner. What next?"

§

As a teacher, I always tried to draw out the best in each individual student. While we might be reading William Faulkner one semester and Maya Angelou the next, debating capitalism one term and vivisection the next, I wasn't satisfied that the young people in my charge would be better writers, speakers, readers, or thinkers when they left me each June. No: I wanted them to be bigger people. If they were shy, I wanted them confident. If they were cynical or narrow-minded, I wanted them stretched and challenged. It was in the keeping of a Buddy journal that I consciously realized that I was approaching animals the same way I had my students. Just whom could each of these broken animals become if all the conditions were right? Just whom? It was certainly true that my dog Murphy and I had a rapport that defied convention. From the time he was a young pup, I'd wondered how much this dog could learn and how well he and I could "talk." I approached our relationship the way I approached my students: believing in his potential and in my ability to draw it out. Our relationship had proven that if one teaches well, watches an animal for cues, and refines communication based on those cues, the animal will far surpass conventional expectations of his species, his interaction becoming increasingly sophisticated, complex, and nuanced.

Murphy's vocabulary, for instance, includes the following words and phrases, among dozens more: carrot, treat, ice cream, friend (his word for dog), cat, deer, big birds (wild turkeys), chow time, hot, drop it, stick, get the ball, get the toy, get the towel, find my shoes, find David, find a stick, find the bird, I need your help, go tell David, show me what you want to do (leads me to the door, his food dish, or to the place where something he wants is inaccessible). If he wants to play, he picks up the toy he wants me to play with. If he's tired and I ask "Want to go to sleep?" he plods into the bedroom and hops onto the bed. If he's smelly, I say, "Murphy, we've got to take a shower," and the sweet, good dog walks to the bathroom and into the shower. His very specific behavior also indicates that he understands "Got to check the animals," "Got to clean your ears," "Give me your feet," "Let's go swimming," "May I have a kiss?" and so on. Since he obviously understands my language far better than I do his, Murphy has also devised a repertoire of sounds, physical postures, and gestures to communicate what's on his mind, including nudging, leading me to what he wants, blocking my path, blinking, licking his nose ("I really want a treat, Mom"), placing his head on my leg, grabbing my shirt (if I've ignored more subtle requests to play), and sounds ranging from a pitiful and barely audible whimper ("I really need you to come down here and give me a hug") to a single sharp bark ("I'm outside the door and I want to come in right now" or "I can't reach my toy"). When I think of animal "experts" who, despite so much evidence to the contrary, argue that animals don't think but instead are entirely instinct-driven, I think of our animals, and want to invite those experts for a visit. Murphy would have a thing or two to tell them.

But there are fundamental differences between a relationship with an animal one has had for the animal's entire life and one with a newly arrived and frightened creature. In his ten years of life, Murphy has had about eight minutes of sadness—all of them when he was caught in an outrageous food theft and was unable to scarf his loot. Would the same strategy work with the broken spirits who were arriving with increasing frequency at our young haven? I wasn't sure. I'd always believed in the healing power of love, in my capacity to quickly know the student before me and to draw him out. In this case, the strategy was the only one I had, and Buddy would be my first equine charge.

So there he was standing in the middle of the driveway, ten feet outside his stall door. He'd found the peppermints, so it was time to do something.

Would he walk with me? I wondered. "Hey, sweetness, want to go for a walk?" I asked.

Well, duh! Why else would this animal have ventured out on his own? Standing on his left, I reached under Buddy's chin for his halter. Though I grabbed a lead rope in case I needed it, I sensed there would be little need for one. With my hand on his halter, I could guide his every step, and, I hoped, more quickly build trust and confidence.

Buddy and I walked about two hundred feet—tentatively, but without a single stop. His head was turned slightly to the left as though willing his right (and only) eye, though stone blind, to see in front of him. He was certainly also listening with great intent. And on that first day, Buddy heard the goats.

Now if you're not fortunate enough to know them, goats are remarkably friendly animals and are insatiably curious about newcomers of any species. Until proven wrong, goats behave as if the earth and its inhabitants are all sources of love, fun, adventure, or food. Yes, theirs is an optimistic worldview. Among abused and neglected animals, they are the quickest to trust. Crotchety old billy goats are sometimes the rare exceptions; we had none of those.

So Walter, Spanky, Frodo, and Ralph trotted over to greet us. Gorgeous Walter, jet black with the basset-hound ears of a Nubian, bleated "hello." All four poked their little noses through the fence. "Who are you?" they asked of Buddy. I didn't matter to them. They were interested in the newcomer, and this both impressed and touched me.

Whether Buddy had known goats in a previous life or simply sensed the amiability of these four animals, I don't know. Whichever it was, his comfort and curiosity matched theirs. The goats huddled as close to him as the fence would allow. Walter and Frodo jumped up and placed their hooves on the top rail to get even closer. Too short to reach, the other two stood their ground, doing their damndest to stretch through the fence to smell Buddy's knees.

He bent to reach the goats. Lots of sniffing ensued, then a little gentle nibbling of Walter's face. To my delight, Buddy was every bit as interested in the goats as they were in him. As we walked away, Walter and company hugged the fence and followed along. Buddy was reluctant to move, but this time, his reluctance wasn't about fear; it was about leaving his newfound friends. Aaah! That he could be eager for animal companionship hadn't occurred to me until now.

We rounded the far end of the oval drive. As we headed back to the barn, the pony paddock was on our right. It was inhabited by tiny, ancient Dino and three other pony friends.

Dino was nearly forty years old. When we'd opened in the winter of 2001, he was our first animal. The sole survivor of a case of arson that took the lives of twenty or more horses, Dino was barely three feet tall, yet he had survived by kicking his stall door down after the flaming ceiling had collapsed on top of him and was burning him alive. He was one brave boy.

At Catskill Animal Sanctuary, though his life-threatening burns and a partially-collapsed lung had healed, Dino's psyche had not. Day and night, Dino stood alone in his pasture, indifferent to his pasture mates. An animal communicator who claimed to know nothing about Dino said she sensed he was wondering where his friends were. Whether she truly sensed this or had simply read our Web site didn't matter. The fire that had killed twenty other horses had left its mark on this little guy. We humans loved him, groomed him, gave him what few treats were safe—mostly finely chopped apples and carrots, as age had taken nearly all of his chewing teeth. We talked, we sang, we brushed, we kissed. Dino stood passively, his head lowered. Perhaps I was projecting, but Dino seemed remarkably introspective. Nothing we did brought him out of himself. No special treat elicited a whinny, none of the other ponies elicited interest. The fire, it seemed, had taken not only his friends but also his spirit. Perhaps he was even grieving.

A gruff, unfamiliar nicker filled the air. Buddy and I were walking past the pony pasture, and Dino was trotting toward us, whinnying his heart out! This is what I saw in that moment:

Dino is excited to see Buddy!

Dino is in far more pain than we realized.

Maybe these two can heal each other.

"Hello, big horse! Who are you?!" Dino limped on ancient legs, his shoulder so arthritic that diets, supplements, massage, and other alternative treatments had little effect. But he trotted through his pain, his head high and his Tina Turner mane bouncing, for when he saw Buddy, he saw a friend. Buddy tugged me toward the whinny.

Was it Buddy's spirit? Buddy's need? Buddy's innocence? Other horses had walked past Dino's paddock and he'd largely ignored them. Tim and Jim, a pair of 18-hand draft horses headed to slaughter until we intervened; Bucky, a broken-down Standardbred racehorse we nicknamed Mr. Lips for his habit of mouthing any body part before him; Musty and Booker, two horses surrendered by their owner so she could avoid cruelty charges.

As soon as Buddy and Dino met, they loved each other. I can't explain it any better than that, except to say that those whose hearts are open probably know this experience. Whether it's been with a romantic partner, a platonic friend, or an animal, most of us have had relationships in which the connection was instantaneous, the feelings clear and irrefutable. Well, on this wonderful day, two old horse hearts were wide open, and one of the deepest friendships I've ever witnessed began that moment, as noses touched between the top two rails of a split-rail fence.

- 10 -

Buddy's Big Day

Ready for an adventure, Bud?" I asked the next morning as I loosened the dust from his rump with a rubber curry comb. In just a week, Buddy had gained forty pounds, and the sheen was returning to his dull coat. He did the usual: he scratched his forehead on my back, and with the last downward movement, lifted his muzzle, and sent me sailing.

"Silly boy, what are you doing?!" I exclaimed. I backed up to him, and we repeated the game, Buddy flinging me forward each time.

A prudent horseperson would probably discourage this behavior. "Your horse must always know that you're the alpha," such people say, and they draw their boundaries tightly. Hogwash, I say to them. These games that Buddy instigated and I allowed—indeed, encouraged—were definitely indulgent, but I considered it part of my job to draw the spirit out of this animal. I loved the fact that in just a week's time, we were connecting as we did. In fact, the longer I do this work, the less I tolerate the "tight boundaries" approach to working with animals. Much of the joy is in bringing out each animal's uniqueness, in encouraging the spirit and play and personality to emerge, in building a relationship of trust and respect that allows both human and animal to move beyond the conventional arrangement in which the docile animal obeys the human upon whom he depends for food and shelter. This is what I do, and it's what I expect of my staff. We don't take foolish risks, but we routinely push the envelope, and are often delighted with the results.

On Buddy's third day out with his seeing-eye human, it was clear that he was ready for a challenge. We headed for the woods. Neither of us had been there before.

The trails behind the barn were nothing like the pristine trails of a state park or other designated hiking area. They began with a fairly steep, curved descent to a small creek. The creek itself was only six feet wide and a foot deep, but its unstable rock bed would challenge even a sighted horse. On the far side of the creek, deep, choppy ruts and a tangle of small trees had been left by a bulldozer. Behind this obstruction, a sixty-degree climb of about fifteen feet led to easier terrain from that point onward, but the several hundred acres of woods and open meadow behind the farm also had their challenges: swamp, steep climbs, dense growth. Yes, I had a feeling my new friend and I would remember this day.

"Down, down, down," I instructed Buddy with every few steps down the first hill. My right hand was on his halter; my left reached across my chest and rested on his neck as reassurance. He moved with confidence; his head was high, his ears forward. A few steps before the creek, I commanded "Stop!" and Buddy did so. We took small steps to the creek. There was no choice; I stepped in first, glad that it was a warm summer day.

"Water," I told my friend. "Water." I splashed and swirled the cool water with my hand. Buddy stood on the bank, listening and summoning his courage, as I backed deeper into the water, my voice projecting confidence and calm. I stood in the middle of the creek; the lead rope stretched out between us. "You can do this, Bud! We're having an adventure!" I praised.

In an instant he was coming toward me. He marched through the creek; I reeled around to accompany him, and commanded "Up, up!" on the other side of the bank. "Good boy, Bud! You walked through the water!" I exclaimed as I steered him to the left to avoid the first rut, and then in a half-circle to bend around the second. Now the steep bank: could we make a sixty-degree climb without slipping backwards? Horses are wonderful climbers; I wasn't so sure my boot treads would hold. But I wasn't about to turn back: Buddy was having too many victories.

"BIG up, up!" I said, and we moved as a team up the steep bank as if we'd done it a hundred times before.

We were just three hundred yards from the barn, but what a walk we had just taken. My heart in my throat, we stopped so I could praise Buddy. The exuberance with which a blind horse—afraid to move one week earlier—had just walked tough and unfamiliar ground, the leap of faith he'd taken in entrusting his safety to me, and the thrill of working as a team all coalesced as I whispered how proud I was of my friend. Buddy wasn't interested in my praise, however. He was searching for a peppermint and anxious to continue the adventure. Yes . . . this boy was having fun, and he was proud of himself! It was in this moment that two things became clear: Buddy desperately wanted to be challenged, and would do anything asked of him.

The next quarter mile was comprised of gently rolling terrain: a smooth trail cut through a forest of mixed evergreen and deciduous trees—maple, oak, birch, and poplar. Buddy lifted his feet high as we traversed unfamiliar turf. How clever he was to do this as a means of avoiding tripping over fallen trees, rocks, etc. We began up a gentle slope; at its crest, a downed tree blocked our path.

"Want to trot, boy?" I asked, clucking, and he instantly responded. Buddy trotted the hundred feet or so to the top as I ran beside him, and a few steps before the fallen tree, I once again snapped "Stop!"

Buddy stopped on a dime. I was breathing hard and thankful for the fallen pine tree. It was a small tree, just six inches in diameter, and it rested twelve or fourteen inches off the ground. I inched Buddy up until his lowered head would touch the tree, and then I smacked the tree several times as hard as I could. "Touch it, Buddy," I suggested. He didn't know the words, of course, but the eager and adventurous horse did know that we were solving a problem together. I smacked the tree again, and he lowered his head a bit too eagerly and bonked right into it.

"Good job, smart boy!" I laughed.

Okay. He had a sense of how high he had to step to make it over the tree. But would he actually do this? I turned him away and walked backwards thirty feet in order to turn again and come at the obstacle with some momentum. Ten steps, now six . . . I edged over so my body touched his shoulder, and held his head closely. Four steps, now three, now two, now one and then "Up, up, Buddy!"

Jesse the calf kisses Kathy. Calves love to affectionately lick their humans.

Rambo on patrol.

Hannah, caught in a rare moment away from Rambo.

"Have you seen Rambo around here?"

Buddy in classic "Buddy pose."

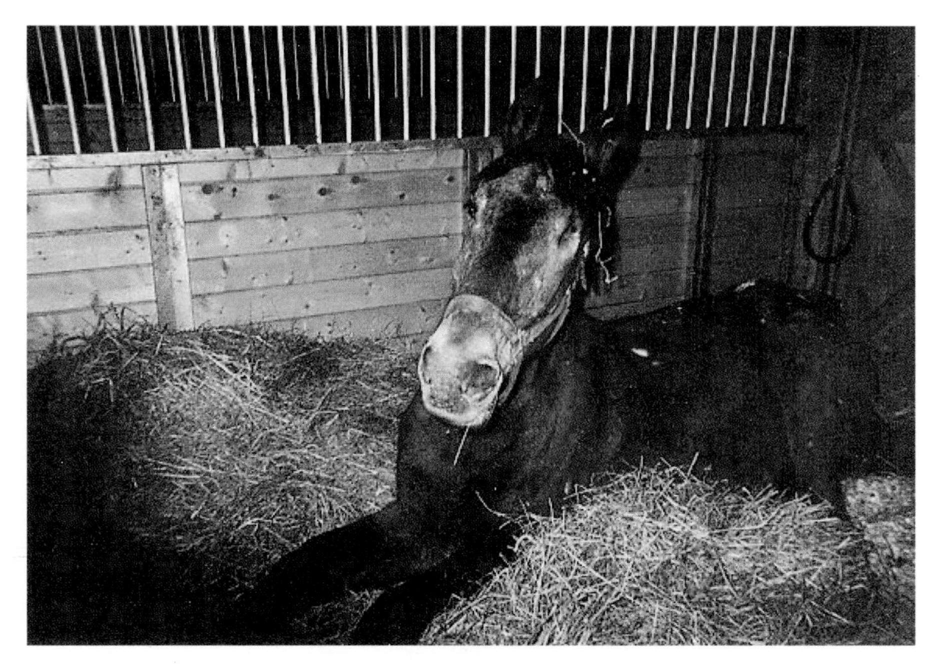

Buddy naps in his stall.

Bobo, another blind horse, has a full, happy life at CAS.

Dino, one brave little guy.

Ted and Dino.

Steve Rucano, of Secure Construction, works on the new cow barn.

Tiano Excavation installs drainage to prevent spring flooding.

Dino and Julie.

April with special-needs cow Rosie.

Board member Chris Seeholzer helps Leonard H. Nelson
reach willow leaves, his favorite treat.

Lorraine with Babe, 2,000 pounds of love.

Lissa and Walt.

Alex with old man Maxx.

Our beloved Paulie.

He lifted his right leg but not quite high enough. The top of his hoof whacked the tree; he placed his leg on the ground. Not a problem. "If at first you don't succeed," some internal voice was saying to the horse, who without prompting, lowered his head to once again feel the height of the tree.

"Up, up, Bud! You can do it!" Buddy lifted his right leg, cleared the tree, and proceeded with the other three. Once all four legs were on solid ground, I turned my back and leaned into his head so the happy animal could send me sailing, which he did not once, not twice, but six or seven times.

Our trek continued for nearly two hours. Along the way we encountered a swamp ("water," I explained), a boardwalk that someone had built over a section of it, a narrow path along a steep bank, many more downed trees. Buddy approached each obstacle with courage, intelligence, and complete faith in me to communicate what he needed to know to navigate it successfully. On the return home, he managed the sixty-degree downhill, the bulldozed ruts, and the creek without a moment's hesitation, and we trotted victoriously up the final hill to the barn. My lungs hurt, my equine partner and I were both sweating, but my heart was swelling with pride for an animal who just a week earlier had given up because the family that loved him didn't understand what he needed.

A wash stall—a 12' by 12' space with a concrete floor, a drain, and a roof overhead—was tucked into a corner behind the barn. A tack box containing shampoo, scrub brushes, sweat blade for flinging off excess water, and towels was in the corner. Hmmmm, I wondered, and then decided our adventure should continue. Buddy was hot, and there was only one way to know if he'd enjoy a bath. "Come on, boy," I instructed.

The concrete pad was a good eight inches off the ground, but in the woods he'd mastered the "stop, touch it, up! Up!" routine almost instantly. He felt where the pad was and stepped up onto it. Rather than force Buddy to stand by hooking both sides of his halter to two long chains called "cross ties," I decided to hold him with one hand, spray with the other. If Buddy was frightened, I didn't want him to be more frightened by being unable to move.

I also didn't think his bath experience should be with ice-cold water, so I turned on the faucet and let the water run for a moment as I adjusted the temperature. I squeezed the nozzle just enough to release a slow stream onto the concrete near Buddy's front hooves.

"Water, Bud. Water, water." His ears were up and his head was high, as it always was when he was trying to figure something out. I directed the spray toward his hooves. He didn't flinch. I squeezed to increase the water pressure, and moved the stream up his legs. Buddy stood motionless. I squeezed again, water surged from the hose, and I directed it to Buddy's chest. The lid of his one eye began to droop: Buddy was enjoying this! As the spray wetted down one section of Buddy's coat at a time, he relaxed more deeply with each increasing minute. I squirted shampoo onto his dripping body, and gave Buddy a good, long bath. The longer I sprayed and massaged with the scrub brush, the more he let go. Much to my delight, he even enjoyed having his head sprayed, something few horses tolerate. Buddy simply closed his blind eye and stood in ecstasy as the spray massaged his entire face, rinsing off the sweat from his woodland trek.

As I dried Buddy off and took him for a brief walk and graze before returning him to his stall, I wondered if tomorrow would be too soon to ride him. He'd only been with us for one week. But I felt certain that if I could have asked him, Buddy's response would have been, "Let's do it!"

Perhaps we would.

Part Two

The Journey Continues

- 11 -

The Doctor Is In

It's mid-afternoon when I walk to the barn. Walt is wearing his distinctive long face and waiting for me to ask him what's wrong. So I glance at the clock and brace myself.

"Oh god, Kathy," he begins. "Everything's gone wrong today. *Unfortunately*, I didn't accurately calculate how long the shavings delivery would last since we're now using shavings instead of bedding hay for the goats. So I ordered an extra delivery of shavings, but it didn't come until almost one o'clock so that set the barn back by *half a damn day*. Then Winston [a little pig] found a new way into the cow field and it took *thirty minutes* to chase him out so he wouldn't become a *football* out there. *Then*" (Walt is wound up now and starting to bounce a little) "the vet was supposed to bring us three bottles of Banamine but forgot, so—like I really have time to do this—I had to drive to f★★★ing *Rhinebeck* because of course you would *kill* me if we had an emergency and didn't have any on hand. The *icing* on the *cake* . . . "

"Good morning, Walt," I interrupt him. That's our cue for Walt to calm down, take a breath, and get to the point.

Unfortunately, he usually just talks faster.

But today it works.

"Okay. Okay . . . I'll just say that today has been a *tsunami* of *dipshittedness*!"

Meet Walt Batycki, resident drama queen, wind-up toy, stand-up comic, and the best animal-care director one could conjure up.

Before migrating east to be near family as he awaited the birth of his son, Walt was an animator for DreamWorks and Disney. At Disney he worked on creative development for the theme park; at DreamWorks he headed a team of animators that developed computer games. "I'll have you know," he points out, "that headhunters were fighting over me." After four years of working with the man, I've no doubt he could have had any job that he wanted.

Walt's final project was to animate a game called "Zoo Tycoon," in which his character's job was to care for the animals, including the dinosaurs that inhabited this particular zoo. He laughs, "So I had plenty of experience to prepare me for this work."

Managing the diets and health of 175 animals from twelve different species is a tall order—even for a zoo tycoon. Walt organizes his time using two lists, the first of which details daily "critical care" issues in need of immediate treatment. Nine animals are on today's list, including a new horse with a hoof condition called "thrush," four pigs with a skin condition called "greasy pig" that arises when the weather is damp, a just-arrived duck found on the side of the road who is infested with maggots, and Rosie the cow who needs a booster shot to end her round of vaccinations. Today's list is manageable: whenever we take in a large rescue, critical care can take the entire day.

The second list organizes routine health maintenance for every animal on the place: a sort of "eyes, ears, nose, throat, skin, superficial wounds" once over. As Walt explains, he "puts his hands on every animal" twice a week, and then schedules into his calendar vaccinations; hoof, tusk, and nail trimming; and de-worming.

Now if Walt were in his former life, he could be a zoo tycoon who skips around feeding grateful goats and contented cows in an otherwise uneventful day. But CAS is a rescue organization—a place where one quickly learns to predict the unpredictable. Our hay man brings 750 bales of hay but no one to help unload it. We hear the squeal of tires at the top of the driveway and discover that sixteen bloody roosters have been dumped. An injured swan appears in our pond; a terrified pig has taken up residence in a young couple's flower bed after his abusive owner kicked him one too many times. Walt has learned to roll with the punches, come up smiling, and transfer today's routine tasks to tomorrow's list.

Even though he claims that coming to CAS was like "coming to Jurassic Park," today, thank goodness, Walt is only working with rabbits and sheep. A sweet gray rabbit named Bailey is cradled in Walt's left arm, having recently returned from an unsuccessful adoption. (Any violation of our adoption contract results in an animal being removed from the home and brought back to CAS. Fortunately, nearly all our adoptions are successful. Bailey's wasn't.)

"He's a horrible mess," Walt laments. "I really can't believe it . . . I was so sure his family would take great care of him . . ." his voice tapers off as he stretches Bailey's lids to examine his eyes. "It will be easy to nuke the respiratory infection and the fleas, but he's also got a nasty abscess and his ducts are blocked."

"Ducts?" I ask.

"Rabbits have ducts that connect their eyes and nose," Walt instructs. "When there's fluid build-up in the eyes, the excess drains out the nose, but his ducts are totally blocked; in fact, they're probably scarred shut. I've sent him to Dr. Rosenberg twice to have them drained, but it hasn't helped."

He coos to the little rabbit as he gently lifts his gums. "Oooh, you're the best boy in the world." I laugh at Bailey's huge Bugs Bunny incisors—two on top, two on bottom—used to gnaw through obstacles and pick things up. "Well, at least the abscess seems to be shrinking. That's good news."

I ask what the implications are regarding Bailey's blocked ducts, and Walt doesn't miss a beat. "There is definitely a question about his quality of life. But I have a plan if you'll okay the expenses."

"Shoot."

"He needs five days without medication, after which he should go back to the vet for a culture that will determine what bacteria is causing the build up. It's possible that keeping him on a really low-maintenance dose of something will keep the trouble at bay."

"Go for it," I say.

Walt places Bailey back in his shelter, but not before kissing him one last time.

He takes a seat to update the medical log book. "By the way, Kathy," he says. "Honey Bunny got his last dose of Ivermec today. Isn't that great?"

Both ancient and fragile, Honey Bunny is far more susceptible to parasites and sickness than the rest of the rabbits. Fur and ear mites appeared suddenly and for the first time ever last week; Walt, ever vigilant, spotted them.

"Bless his old heart," I say.

I wait for Walt to remember why I'm here. But he is all motion now, moving toward Paulie the rooster's stall in order to check his ears. Since he came to live with us three years ago, Paulie has been prone to severe infections in his left ear; we speculate that years of cockfighting damaged it—perhaps there's even a tiny tear or missing bit on the feather flap that serves to keep out dirt and other foreign bodies. A crust on the feather flap precedes a full-blown infection, so daily checks are an essential preventative practice. Paulie certainly doesn't mind. The more attention he gets, the happier the old rooster is.

The clock is ticking, so I nudge a little. "Hey, Walt . . . ready to do Noel?"

He stops in his tracks. "Oh, yippee. What a beautiful day in the neighborhood, Mr. Rogers!"

Walt wormed and trimmed the hooves on our flock of twelve sheep several days ago. They're friendly and accustomed to the routine, so the two procedures on all twelve take just over an hour. But then there's Noel.

A Barbados sheep, Noel is one of hundreds of animals to come to CAS via New York City. Henry the rooster was found in a Bronx mailbox, a young goat named Oliver was found wandering Manhattan streets with SOLD spray-painted on his side and what looked like a knife wound on his neck. Hannah the sheep was found in a Queens cemetery. When Animal Care and Control called on Christmas Eve about a sheep being chased down a Bronx street by a pack of dogs, we didn't even blink. "Sure, we have room," we told them.

Little Noel was wild-eyed with terror. No matter how slowly and quietly we entered her stall, she leapt up walls in a vain attempt to escape. Nearly a year later, though she'll approach us with caution at mealtime, she is still terrified of handling, so even routine health care is arduous and exhausting.

"I'll rally the crew if you want to get your supplies."

"Okay, sure," Walt agrees.

Five minutes later, six of us head to Noel's field: a small pasture behind my house where she lives with two goats named Denzel and Damon and two sheep named Christopher and Jack. This is our "special needs" sheep and goat field, where Noel lives not because she technically has special needs, but because we need to catch her—a feat that would be impossible in the much larger and hilly sheep pasture.

Noel, it seems, is slowly getting better. Though we all work up an appetite, it takes us only ten minutes to corner her. Walt gently loops the lead rope around her neck, and the gang returns to barn chores.

He lays the sheep on her side in the cool grass and I lean on her to keep her still. I've never felt a heart beat so fast. Walt squirts de-wormer down her throat before she can fuss, and then, as he moves expertly from one hoof to the next, trimming off the excess, he serenades the frightened girl with "Noel the Crazy Girl Sheep," sung to the tune of "Puff the Magic Dragon":

Noel, the crazy girl sheep
Lives on the farm
And frolics with Denzel and Jack
We'll never do her harm . . .

"You know," he jokes, "I left them the instructional video *What to Expect From Your Hoof-Trimming Experience* last night. Noel obviously didn't watch it."

§

It would certainly be a lot easier for Walt if the animals spoke our language, or we theirs. A limp can mean so many things: an object lodged in the hoof; an abscess; a sprain, tear, or bruise in the hoof, leg, or shoulder. It can indicate the beginning or worsening of arthritis, or myriad hoof conditions—and these are just for starters. How great it would be if Mr. Specks the goat could say, "Walt, Austin knocked me off the rocks and I hurt my leg." But we don't share the same language, and Walt's only diagnostic tools are a thermometer, a stethoscope, and his own eyes, ears, and intuition.

He can't even use language to reassure animals that he is the same kind, gentle Walt on health-check days that he is every other day of the year. Even though his gentle rendition of "Noel the Crazy Girl Sheep" probably soothes her on some level, Noel is still terrified. And so is Babe, the 2,200-pound steer, whenever he sees anything in Walt's hand that's not a bale of hay. Soothing words do not work, because Babe can't understand them. On cow health days, a syringe, a bottle—even, for goodness sake, a wad of warm cotton for ear cleaning—is a deadly weapon, and Walt is Darth Vader sent to destroy him.

In his four years here, Walt has developed an encyclopedic knowledge of farm-animal ailments and their symptoms. He has become an exceptional caretaker and a crackerjack diagnostician. Significantly, his skin has also thickened a little—it had to. Walt has had to learn not to take it personally when Babe runs from him for several days after he's been vaccinated, or when Claude the pig gives him the cold shoulder for a solid week after Walt has trimmed his tusks. And this is the easy stuff. Far more importantly, he's learned that one can't always wait for an animal to know that it's time to go, and that sometimes we have to make that toughest of calls. It's in these moments that I'm proudest of him, for he's learned to leave the dramatic Walt at home. He quiets himself for the dying animal, comforting him with every cell of his being, and then, when the animal is gone, he cries with the rest of us.

Four years ago, Walt couldn't have distinguished between a pitchfork and a rake. But today, we are blessed to have this gentle man in our midst.

"I don't want anyone to suffer, and I especially don't want anyone to suffer because I didn't do something," Walt explains. He does what it takes, every single day, even when it means working for fourteen hours. The reason is simple.

"This place makes my heart sing," Walt explains. "These animals are gifts from God—like getting a thank you note from heaven that says, 'You're doing the right thing.'"

-12 -

Every Day is Pigs' Day

As exasperating as they can be, we at CAS love our fourteen pigs madly. But as they taught us well and quickly, working with the porcine species is quite unlike working with any other farm animal. In the human/pig relationship, the human adjusts to the pig—period. Unless one is willing to beat or terrify or confine them into submission—precisely what happens in the factories that grow the vast majority of our pork—pigs simply won't have it any other way. Go blindly into a relationship with a pig, and you'll very likely come out enraged, injured, and deeply disgraced.

So, in honor of this delightful, maddening species (and to spare injury and insult to a human here and there) here are the lessons we've learned, masterfully shared with us by Police, Rosie, Jangles, Charlie, Claude, Babe, Zoey, and the rest of the CAS pig posse. (A disclaimer: We are neither stupid nor masochistic. Much of what you're about to read happened when we were pig neophytes.)

§

LESSON ONE: IT AIN'T EASY TO OUTSMART A PIG

On lists that rank species by intelligence, pigs are generally listed fifth: humans, primates, dolphins, elephants, pigs. Without a doubt, they are sometimes so smart that one can feel humbled (if not utterly humiliated) at unsuccessful attempts to outwit a porcine pal. Just ask Alex how many new locks he's devised to keep them out of the kitchen. Yes, our relative positions on those intelligence lists are questionable.

§

LESSON TWO: CAN YOU SAY "DRAMA QUEEN"?

Prior to Petunia's arrival, it was only anecdotally that we knew of pigs' proclivity toward melodrama. So it was baptism by fire when she arrived one summer day in 2003 and refused to leave her trailer.

"She's gonna scream," her surrenderers offered in the way of help. But they didn't offer to drive closer to the barn or to take her in themselves. They had no harness, so I looped a lead rope under her neck, thinking I could simply encourage her with words and gentle tugs.

Apparently it was time for Lesson Two: It doesn't take much to make a pig hysterical. On a single exhale, Petunia's deep growls

Oreo runs through the grass.

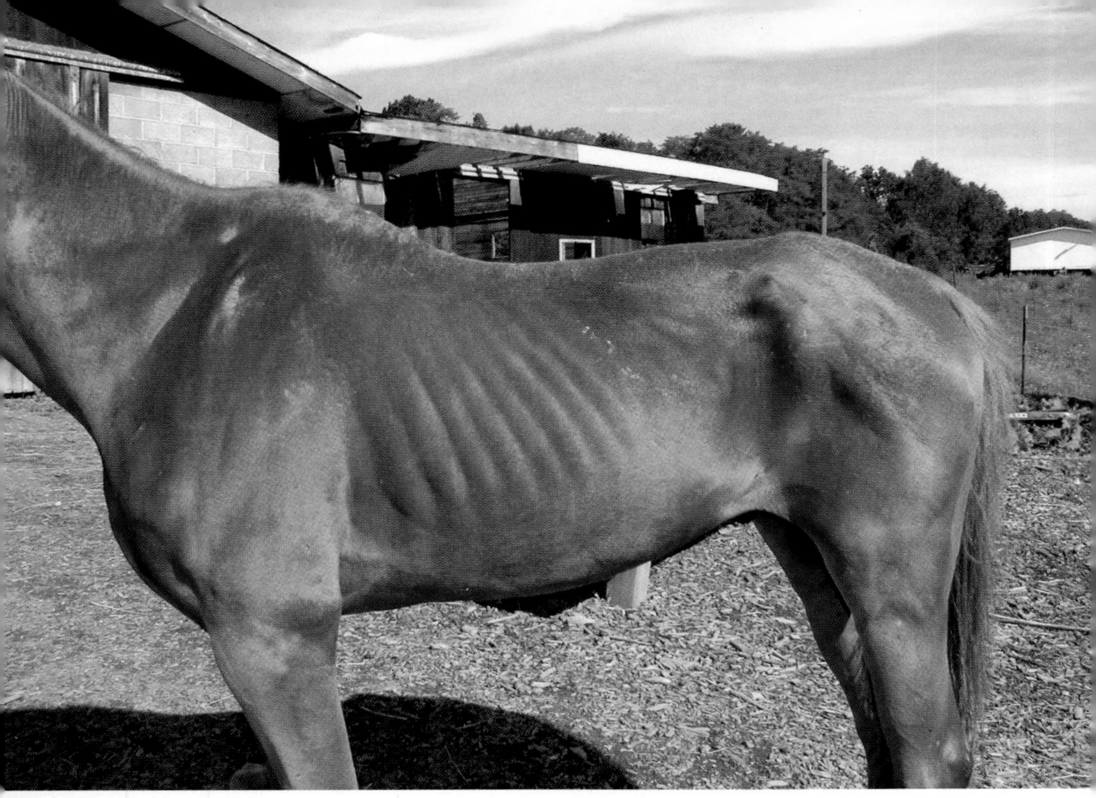

Beyond, when he arrived.

One of 300 chickens abandoned at a live poultry market in Brooklyn.
They were stuffed approximately thirty birds to each 2' x 2' crate.

Oliver, found wandering Manhattan with "sold" spray-painted on his side.

Beyond, rehabilitated.

Christopher, just a few hours old.

Just a little sleepy.

Franklin says, "I want a pumpkin!"

"But, okay, a squash will do."

Policeman naps with a friend.

Zoey napping.

Murphy supervising barn chores.

The great dog Murphy, Director of Canine Pursuits.

Have you hugged a goat today?

"Give me a kiss!"

Walt greets Misha.

Three Pekins relax in our duck area,
a spacious habitat enclosed to keep out predators.

Samson

Samson, forever memorialized in a life-size painting.
The words "Peace to All Who Enter Here" are engraved above him.

"Hey, good lookin'!"

became a deafening, high-pitched, "I know you're trying to kill me!" scream, and in an instant, I was flat on the ground, being dragged down the driveway by a pig whom I could neither control nor comfort. I let go.

So hysterical are they when nervous that Walt no longer schedules pig health checks on visiting days. "The dirty looks got old," he explains. When Walt cleans the pigs' ears, they grumble. When he trims their hooves, they growl. On tusk-trimming day, get out the ear protection. Mind you, none of these procedures hurt: trimming a hoof or a tusk is little different from filing a fingernail. But try telling that to a pig as he's flying off the handle, or to a guest as she's convinced that Walt really is torturing the poor pig.

"I hate it when guests stand there looking at me like I'm Attila the Hun!" Walt complains. "No matter how I explain that what I'm doing doesn't hurt at all, and that pigs are just hyper-emotional, the snide remarks always come."

"Like what?" I ask.

"Oh, like my favorite of all time, 'Young man, there's no need for that. No need.'"

§

LESSON THREE: PIGS CRY

Not long after Petunia arrived, she claimed CAS as her fiefdom. She roamed the farm searching for food and friends (mostly food), dropped to the ground in front of us to demand belly rubs, used a bench to hoist herself up to the table to steal volunteers' lunches. As she ruled over Fortune Valley, Petunia was one happy pig.

But one day, trotting in from his pasture, Austin the goat butted Petunia: one swift, impulsive blow with a colossal horn. The blow failed to nick Petunia's tough skin, but Petunia shrieked, charged to her stall, buried herself under her hay and wept.

"What do you mean she was crying?" I asked Charlotte Mollo a few minutes after the incident. Charlotte and her husband Walter McGrath have volunteered at CAS since 2001 and had a soft spot for Petunia.

Charlotte touched my arm. "Kathy, she was sobbing. It was indistinguishable from human crying. Lorraine and I saw what happened, we went in to comfort her, and she was inconsolable."

Charlotte and Lorraine lay down in the soft hay with Petunia. One woman hugged Petunia close; the other stroked her face. Petunia wept: it's the only way the two can describe what they witnessed.

"Think about it," Lorraine says. "CAS was her haven, her safe space, a place where nothing bad had ever happened to her, and Austin came out of nowhere and attacked her in that sacred space. She was shattered."

Several influential studies have documented the similarities between human and pig emotions. Noted autistic scientist Dr. Temple Grandin, author of the memoir *Thinking in Pictures*, suggests a biological explanation. In her first human anatomy class, she was "astounded to learn that the limbic system, which is the part of the brain associated with emotion, looks almost exactly like the limbic system of a pig's brain."

§

Lesson Four: Pigs Are Willful (That Might be the Understatement of the Century . . .)

"Willful" is the word I choose. Others choose stubborn, determined, obstinate, pig-headed. Yes, pigs are pig-headed. What else would they be? Whatever adjective one selects, there is rarely a sort-of in a pig's world. It's not often that a pig sort of wants to do anything. She either desires to do or not do something with every fiber of her being . . . and often gets her way.

Take today, for instance. Franklin is angry. At six months old, he's finally become WAY too difficult to manage as a free-range pig. Yesterday, he broke into one of the chicken yards and ransacked the coop, gobbling up eggs as if he hadn't eaten in a week, screaming like a banshee when we kicked him out, and then again when Alex placed him in time out (his cozy stall) for the fourth time that day. In theory, as much as we'd love for Franklin to be a permanent part of the Underfoot family, he is a 250-pound mischief maker: an exasperatingly willful, exceedingly bright, unrelenting whirling dervish. Franklin is all pig, all the time.

So today, in the field right next to the barn in full view of all the goings-on, Franklin has been turned out with the goats and is none too pleased. In fact, he is so intent on indicating his displeasure that a) we all need ear plugs and b) April is a nanosecond from letting him out because she can't bear the thought of our little man being unhappy.

Franklin knows this, you see. He knows that if he acts insane for a few minutes, one of us will give in, the way we always do. Out there in the rest of the world, pigs have little chance of happiness. Humans believe that we're entitled to use animals for our every need, desire, and whim, with virtually no regard for how the animals might feel about such arrangements, or for their welfare as they're being raised to feed, clothe, or entertain us. But at CAS, our roles are so clearly reversed. We humans are the servile ones.

At the moment, Franklin is pacing the fence, perfecting his "I'm a crazy pig" routine. Pacing, in fact, is the wrong word to describe the behavior. Motionless at one corner of the pasture, he is suddenly a jet-propelled pink blur until he reaches the far corner, where he slams to a halt, erupts like a volcano, then hurls himself back to where he came from. He does this a few times until . . . wait . . . what's that . . . does Franklin smell sympathy?

Though we're hidden from view, somehow the little hellion knows we're watching him. He lifts his snout to the air and sniffs, then, cutting diagonally through the field, runs directly to the gate and waits, "harummphing" loudly, looking toward the barn with the big brown eyes we are always and utterly unable to resist.

"Can I pleeeeeease give him a pumpkin?" volunteer Allen Landes pleads. Allen is a hospital biologist during the week, but devotes his weekends to working tirelessly for CAS. He loves the entire CAS crew, but if pressed, just might whisper that Franklin is his favorite.

"Oh good grief, okay," I say. We're defeated once again. Secretly, I can't wait to watch our little imp bite a hole in the baby pumpkin so that he can race gleefully around the field (the field that just moments earlier was his prison), holding the pumpkin in front of his snout like a bulbous appendage, exclaiming, "Look, world! I have my favorite treat! A pumpkin! IT'S MY FAVORITE!"

§

LESSON FIVE: PIGS WILL HURT EACH OTHER

In general, farm animals will work out their issues with a minimum of conflict. Within a pasture, cows and horses cluster with their particular friends. Occasionally there's an outcast who gets harassed, or a loner who's not allowed to join the herd, but serious conflicts are rare. Not so with pigs. Giving them time and space to work out their issues can easily result in extreme emotional distress and/or injury to vulnerable animals. One farm sanctuary found this out the hard way when, after its director insisted on grouping all pigs together, an elderly pig died of a heart attack after being violently attacked.

From the moment they arrived last winter from a sanctuary that closed its doors, Jangles, Judy, and Farfi acted as if they were coming to claim land granted to them by the king. Far younger than Belle and Policeman, the three immediately formed an impenetrable line blocking access to the 25' by 35' heated barn built to house our eight farm pigs. If either Belle or Policeman approached tentatively, the three younger pigs would lunge, sending the two disconsolate elders back out into the cold. When Judy's tusk (despite the name, Judy is a boy) tore a deep gouge in Belle's cheek, we abandoned the idea of a single dwelling. A full year later, three lucky pigs have nearly a thousand square feet of deluxe accommodations to themselves each and every night.

Indeed, our fourteen pigs are housed in seven different arrangements. Charlie has a studio apartment. Franklin, Belle, and Police share a duplex, with Franklin inhabiting the left unit, the other two the right one. Potbellies Zoey, Hampton, Priscilla, and Rosie reside—usually harmoniously—in a housing co-op. And so on. Why do we go to such trouble? Because Jangles, Judy, and Farfi dislike older, more fragile Belle and Policeman. Because Belle and Policeman love each other and no one else. Because Charlie is a loner. Because Babe tolerates Claude, whom she met as a needy adolescent, but despises the other pigs. Because Franklin . . . well, because Franklin is so annoying that any pig other than St. Policeman would probably kill him. Because, in short, we're unwilling to chance injury, and almost equally unwilling to chance hurt feelings.

§

If I have left any doubt, let me dispel it right now: at Catskill Animal Sanctuary, the pigs run the show. While in theory April runs the barn, when you've got fourteen highly intelligent, extremely emotional, willful beyond imagining, food-obsessed animals (half of whom weigh at least a thousand pounds), every single day of the year is Pigs' Day.

It is difficult to fathom how anyone could subject such creatures to the conditions endured by millions under agribusiness each year. Only a handful of megacorporations control nearly all of the pork production in the United States. While the individual farmer may still own his land, he is under contract with Smithfield or another giant—a giant that controls every aspect of the farmer's operation.

One of the most horrific practices of the industry that turns these delightful animals into bacon, sausage, and pork chops is the use of the "gestation crate": a 2' by 6' space with metal slats for floors and walls of steel bars. These crates (deemed cruel by the European Union and scheduled to be phased out by 2012) are where breeding sows, or mother pigs, spend nearly all of their lives, with absolutely nothing to do except eat food that is mechanically delivered to them. Breeding sows cannot interact with other pigs. They cannot root or forage. Not only can they not exercise; they cannot even turn around. They stand on steel bars, so they have no straw with which to make a nest. Nest making is a task of such significance to pigs that potbelly Charlie often crosses the barn aisle, steals a mouthful of straw from Police and Belle's stall, and takes it back to his own stall, trip after trip. Yet mother pigs, driven by the desire to keep their piglets warm and comfortable, are deprived of the ability to do so. Intelligent, social, emotional animals, animals just like Franklin and Petunia.

Other common practices are equally appalling: castration without anesthesia, tail removal without anesthesia. For farmed animals, legislation regarding their care begins with their transportation to the slaughterhouse. Though woefully inadequate and largely unenforced, at least a few trivial laws exist to regulate transportation and slaughter. Agribusiness does as it pleases while the animal is on the farm—or in the factory, to be more accurate. Its sole concern is to grow the largest number of animals as quickly and efficiently as possible in the smallest amount of space. There's no room in such a scenario for even the tiniest concession to their physical or emotional comfort.

Milan Kundera, author of *The Unbearable Lightness of Being*, wrote, "Humanity's true moral test, its fundamental test, consists of its attitude toward those who are at its mercy: animals." Fellow humans, we're failing that test.

- 13 -

Paulie Comes Home

The Mayor's Alliance for New York City's Animals had invited Catskill Animal Sanctuary to be an affiliate member. No matter that they were a coalition of dog and cat shelters from the five boroughs, they invited us to participate, so we did. I considered the membership a wonderful networking opportunity . . . I just didn't realize I'd be networking with a fighting rooster.

"We're setting up in Prospect Park near the rotary," said Jane Hoffman, president of the Mayor's Alliance, describing the first of the Alliance's Pet Adoption Festivals. "Do you have a tent?"

"Yep."

"Good. Bring some animals."

I smiled to myself. Hauling farm animals around to city events was not only implausible, it would be too stressful on the animal for us to consider it. I appreciated her thought that real animals would generate interest in our work, but brochures, newsletters, photo albums, and reprints of news articles on our rescues were all I'd be bringing. This collection and my own enthusiasm would have to suffice.

"Catskill Animal Sanctuary? You take farm animals?" A striking silver-haired man, Jake, stood in front of my booth. His T-shirt advertised the name of a local dog and cat rescue organization.

"Farm animals only," I said, smiling, and held out my hand. "So don't you try to sneak six cats into my car! I came here alone and I'm going home alone."

"Damn . . . are you sure?" Jake smiled back.

"Pos-i-tive."

"I couldn't talk you into taking a killer rooster?"

That was how Paulie came to live with us.

Paulie, a beautiful russet-colored rooster with forest-green iridescent tail feathers, had evidently been a first-rate cockfighter before finding his way to the dog shelter's basement. In this illegal but popular "sport," two trained roosters with deadly spurs are put into a small enclosure to fight bloody battles until one of the birds succumbs.

Not a particularly large bird, Paulie must have been pure heart. The shelter welcomed him despite his reputation, but then, overrun by dogs and cats and knowing nothing about "killer roosters," shunted Paulie off to the basement. Eventually, they brought him out for Adoption Day at Prospect Park, hoping, I suppose, that among the scores of passersby would be someone looking not for a beagle mix or charming calico but for a former cockfighting rooster with three-inch, razor-sharp spurs.

Evidently, I was to be that person.

Inside an air-conditioned van was a motley crew of canines and felines, each as deserving as the next one of a human all his own. "Hey, pups! Hey kitty cats!" I called. A few tails thumped;

some eyes brightened. I wished hard for all to go home with loving families.

And then I saw Paulie. He was set up inside a fenced area about six feet in diameter and was napping in a corner.

"Now watch this guy," Jake warned. "I'm telling ya . . . he's tough."

No sooner had Jake opened a small gate than Paulie flew right at him, deadly spurs pointing straight at his thigh. Jake threw out his arms to prevent contact. "See what I mean?" he exclaimed. "Look at him! Look at him!"

Well, that's exactly what I did, and I saw not a killer, but an eight-pound bird simply defending himself against someone he perceived as dangerous.

I asked Jake if I could hang out for a few minutes. He told me to be careful, then went to assist people interested in more conventional pets.

Paulie, standing tall in the middle of the pen, soon relaxed, and I crept slowly into his space.

"Hello, beautiful bird!" I cooed. As I've taught the staff to do with new arrivals, I sat on the ground, trying to be as small as I possibly could. Two years earlier, when a stunning pair of frightened steers named Babe and Lincoln arrived, they confirmed what one might know instinctively: the smaller one was, the less fearful the two boys were. Two very young guests arrived one Saturday. Allowed by their trusting parents to enter the pasture with me, the little girl, not yet two, scrunched down into my lap about twelve feet in front of Lincoln. "Come right here, sweetie," I encouraged her brother, Will, as I patted my left knee. The two sat, rapt. "You can help us teach the cows that people are nice and will never hurt them," I explained.

"I want to do that,"Will whispered.

"I know you do, Will! So let's first let them get used to our being here. They need some time to know that we're not going to try to grab them or run at them."

"We just want to be your friends, cows," the angel on my knee uttered.

"Fwiends, cows," the angel in my lap agreed.

We held out our hands to the great beast and his friend. After a few minutes, Lincoln's lids closed slowly over his dark eyes—the same slow, languid blink used by many species to convey love, acknowledgment, warmth, or gratitude. He stepped forward.

"He's . . . coming . . . "Will whispered, barely able to contain himself.

"He sure is, guys. He's coming to say hello."

With a few more steps, the beast who in several months at CAS had not allowed staff or volunteers to touch him, the great black beast with horns the width of a Honda was before us, reaching tentatively but without fear to touch a human hand with his wet nose—the hand of four-year-old Will.

Animals respond to the purity and innocence in young children. Those who have had to fear humans also seem to feel safer when we are physically small. At CAS, crouched and vulnerable and motionless on the ground, we speak softly to the terrified ones, and let them approach us in their own way, on their own terms.

This is how it was with Paulie.

"Hello, beautiful bird!" I began again. "What a handsome boy!" I praised him, slowly squatting into a cross-legged position as I mimicked the "coo" of happy chickens as well as I could.

If there had been more drama in this moment, I'd certainly describe it. But there wasn't, for the rooster who'd first known terror and torture and had later been relegated to a basement for months on end knew in an instant that I was no foe. More importantly, a tiny animal who'd never for a moment been understood apparently understood that I was a friend. At least it seemed so, for the little red bird who moments earlier had gone after Jake came very, very close, then sat down to look at me.

"Well, hello, Paulie!" I exclaimed. "Thank you for coming to see me!" Paulie tilted his head upward in order to make eye contact. "May I pick you up?"

Beneath the fluff of shiny feathers, Paulie was a thin little thing. I hugged him to my belly with my left hand over his wing. How relaxed he was! I slid the thumb and index finger of my right hand under his feathers until I found his tiny neck, which I massaged until, about a minute later, the killer rooster was sound asleep, his tiny head resting in the palm of my hand.

- 14 -

Hey, Teach!

We'd had our share of roosters in the three years prior to Paulie's arrival, but they'd all been "broiler" roosters—birds grown to be slaughtered and served on our dinner plates or in lunchtime sandwiches. In these first years, roughly forty of these birds had escaped their fate and found their way to us. Many were found wandering Manhattan, Brooklyn, or Bronx streets. Dear Henry, one of my favorites, was found in a Bronx mailbox. Ricky and Ruben were tied to a tree in Central Park; young boys were throwing rocks at them. A dozen helpless and bewildered young birds were pulled from the dumpster of a Great Neck condominium.

As rocky as their beginnings were, these birds were fortunate to escape the fate of the billions of chickens grown for meat in the United States alone. Exempt from the Humane Slaughter Act passed by Congress in 1958, "broiler" chickens have no legal protection whatsoever from an industry driven only by profit. In the warehouses in which they are grown, broilers are packed so tightly that they cannot walk around or stretch their wings, as the National Chicken Council's guidelines only require that each chicken of market weight be allowed ninety-six square inches of room—the size of a standard piece of typing paper. As Peter Singer and Jim Mason explain in *The Way We Eat: Why Our Food Choices Matter,* "When the chicks are small, they are not crowded, but as they near market weight, they cover the floor completely—at first glance, it seems as if the shed is carpeted in white." Bred to grow to slaughter weight in just six weeks, broiler chickens experience chronic joint pain for the last portion of their lives, and thus spend most of their time sitting. It is simply too painful to stand. Sores and blisters develop on their breasts and legs from resting in their own excrement—excrement that is typically removed from the buildings once a year or less. Many birds go blind from the ammonia; probably all have severe respiratory difficulty. Heart attacks are so common and so violent that the industry has chosen the term "flip-over syndrome" to describe what happens to the many birds who are found dead on their backs, barely noticed symbols of the breathtaking callousness of agribusiness.

At the slaughterhouse, the only concern is how many dead chickens are ready to ship out each day. There is no concern for the birds; there is no concern for the workers. The assembly line onto which chickens are shackled upside down moves so swiftly that a huge percentage of birds miss both the "stunning" tank (which actually just paralyzes them to make cutting their throats easier) and the throat-cutting stages, and wind up being boiled alive.

Each year, some broilers escape their fate and find their way to sanctuaries. Like their slaughtered brethren, these "Frankenbirds" have been engineered to grow abnormally large breasts abnormally quickly to satisfy consumer demand for more chicken—and specifically, more white meat. So while their relatives are raised in packed, ammonia-filled warehouses, those given a pass grow huge, then huger—quickly. With that growth comes the same problems endemic to human obesity: respiratory issues, heart problems, severe joint pain, arthritis, and greatly shortened life expectancy.

These roosters, morbidly obese and barely able to walk, had been our only experience with roosters. With these birds, their health and well-being was our concern. With Paulie, it was the humans I was concerned about. If anyone was going to be spiked, jabbed, or maimed, it would be me. Besides, I felt that this bird and I had a lot to learn from each other.

So we made a space for Paulie—a big, roomy half-stall, 6' by 12', bedded deeply with hay and located next to the feed room/ kitchen, where Paulie would receive the most attention. It was initially in this space that Paulie taught me, in short order, more than I'd have believed possible about the intelligence, loyalty, curiosity, and courage of chickens.

When they're threatening or warning an interloper, roosters clearly mimic their Spanish friends. One wing stiffens, its feathers spread as the wing is flung to the ground with great flair, and then the feet begin a tango-like stamping, hilarious in its melodrama, until the bird glides sideways toward its target, hiding behind his red cape—I mean, wing. The head tilts, the eye, a perfect pea-sized glinting golden globe, bores into you, daring you to display similarly. Of course you don't, though you're dying to.

My hope on this first day is that Paulie feels not only that he is safe, but that he belongs here. Surely no bird in a tiny cage feels like he belongs in it any more than an imprisoned human would. What does this roomy stall feel like? What does it feel like to be in a barn alive with the sounds and sights and smells of chickens and horses and goats and pigs? Does Paulie know that lots of hens are tucked away safely just twenty feet down the aisle? I wish we could talk right now—I have so many questions! Instead, over the next few glorious weeks, we will observe each other: we will both learn, we will both teach, and we will both be changed by the experience.

We proceed without a plan. Healing physically ill or chronically starved animals requires knowledge and observation. One can easily kill a starving animal by introducing feed too quickly, just as changing too quickly the diet of an animal who has eaten only moldy bagels her whole life can have grave consequences. When the issue is physical—whether it's the consequences of prolonged starvation, dehydration, parasitic infestation, hooves so overgrown that an animal limps, or some combination of these and other hallmarks of chronic neglect—one simply proceeds as needed along a course of care. Symptoms present themselves; one knows through experience what to do.

Treating psychological trauma is different. There is no road map. Unless they've been socialized to trust humans, most farm animals' fear of people develops quickly. Indeed, it's passed from the parents down to each successive generation. When the humans they've known have confined them and repeatedly forced them to fight for their lives, one would think that trust would be a long time coming.

But not with Paulie. He wanted to know us.

"Paulie!" I exclaimed softly as I entered his house. We'd constructed it using lightweight gates made from aluminum tubes, then lined the insides of the gates with chicken wire to keep Paulie in. All day long he could watch as staff, volunteers, and animals came and went.

"Did you have a good night, Mr. Man?" I scrunched down into the soft hay and sat very still. The matador flung down his cape and jerked his neck to look up at me. It wasn't a real threat, however; it was merely an experiment.

"Oh, silly bird, I'm your friend," I cooed as I very delicately slid him away from me with my lower arm. He danced toward me again. Again, I nudged him away. When he came at me a third time, I picked him up gently and pushed my fingers under his neck feathers to reach the tiny, vulnerable neck. Instantly Paulie relaxed. I watched him closely, and again, as he did when we'd met just days before, Paulie slowly closed his lids. If you have cats, you know this behavior well. We've witnessed it in happy horses and contented cows . . . but in chickens? Yes, in chickens! Though chickens use the slow blink to express a wide variety of positive feelings, Paulie right now was simply saying, "This feels really good." The lazy, languid blink was accompanied by a soft little coo, and then, in a moment, by a beak motion like the pursing and then releasing of the lips. All this within a moment of picking up a bird who'd never known kindness was remarkable to me, for Paulie was expressing joy and even, perhaps, gratitude. We had a lot to learn about chickens, it seemed, and this boy was here to help us.

It was not long before Paulie knew his name and would come running when we called him. We began to let him roam free during the morning hours, when we'd lock away Doodle and Scribble, a handsome pair of roosters whom Paulie had determined were his mortal enemies. In the afternoon we'd reverse the routine: Doodle and Scribble would come out to play while Paulie returned to his stall.

"Paulie Bird! I love you, Paulie!" I called and here he came, utterly unable to fly but trying his damndest nonetheless. Paulie accompanied Murphy and me on our walks through the woods, shaking his head and bobbing up and down in my arms at the new sights and smells. I'd bring him to the house, just a hundred feet behind the barn, and as I checked my e-mail and returned phone calls, I'd hear the "click, click, click" of his claws on the floor as he explored the rest of the house. Invariably, though, Paulie would belt out his best cock-a-doodle-doo. Aaaah—it was so obvious—he was calling me! He wanted to know where I was! Could it be true?

"We're in here, Paulie!" I'd call. "We're right here."

The first time this happened, the happy bird appeared in fewer than five seconds. He pecked around for a few moments and then, finding nothing edible, climbed up on Murphy's bed, sidled against the good mutt's belly, and joined him in a nap.

§

One bitterly cold January night, Paulie had more in mind than cozying up to a warm yellow lab. Bundled from head to toe, I trudged to the barn for night check. Everyone was fine—sleeping soundly in their heated or heavily bedded and insulated spaces—until I got to Paulie. Was he shivering? Oh, dear, yes he was! Despite the radiant heater just six feet above his head, Paulie was shivering noticeably, and there was no way to make him appreciably warmer . . . at least not in the barn. He'd have to be an overnight house guest.

I grabbed a crate from storage, filled it with fresh hay, and closed the door tightly. "Come on, boys," I said to my dog and my chicken, "Time for bed." (Good lord, how I love Murphy. Nothing about this life fazes him. Tonight, he would share his home with a chicken, and he was fine with that.)

At the house, I parked Paulie's hay-lined crate in the dining room and filled a dish with water. It was only 9 PM, but I knew I'd be wakened by Paulie's crowing at or even before the first finger of light reached across our valley. "We love you, bird," I said, reaching in one last time to rub his neck, and then I turned to my best pal. "Come on, dog, time to go to bed."

"Cock-a-doo! Cock-a-doo! Cock-a-doo-DOO-DOO-DOO! Cock-a-COME IN HERE RIGHT NOW DAMN DOOOOO!" Murphy and I were awakened from reverie some time before midnight. Paulie was loud and incessant, the meaning of his crowing instantly clear.

"Okay, Paulie, I'll be right there," I called. At the foot of the bed, a patient dog sighed.

I felt my way into the dining room. "I'm right here," I grumbled. Paulie fell silent. Victory was at hand, and he knew it. I picked up the crate and its spoiled tenant and returned to the bedroom. I placed Paulie on the floor at the foot of the bed. "You got what you wanted, P-head. Now go to sleep."

I crawled into bed. "Sorry, Murph," I apologized to the yellow lump. "I love you."

No sooner had my head hit the pillow than Paulie started

up again. "Cock-a-doo! Cock-a-doo! You're a sucker! Murphy is too!" he crowed.

Now I work too hard for my sleep to be interrupted by a squawking anything. I turned on the lights, tore back the sheets, apologized a second time to Murphy. I got down on my knees in front of my former friend.

"What?" I demanded, though what he wanted couldn't have been clearer.

"SQUAWK!"

I opened the crate door. Paulie strutted deliberately to the head of the bed and stretched his neck as high as he could to tell me where he wanted to be.

"Give me a minute, please!" I said as I grabbed a handful of old towels from the closet. I pictured my Dad watching this moment, shaking his head and swearing, "Well I'll be a son of a bitch," underneath his smile.

On the softest pillow I had, I made a nest right beside me for our new bed partner, who quickly made his way across my belly and onto his new bed.

"Good night, bird," I said to him.

"Coo-cooo," he said back, and gave me a slow, appreciative blink.

The next morning, I awoke first. Paulie was right by my head, still snuggled happily into his plush pillow. He hadn't moved an inch.

- 15 -
Lunchtime

"Come on, Murph," I call to the yellow dog who trots down the stairs, a brand new bath towel in his mouth. I don't know why I allow him to do this. It's shortly after noon on a sultry summer Saturday, and I'm walking the two hundred feet between my home and the main barn.

"Bad boy, Franklin!" I hear Lorraine just ahead of me on the left. One border of a large horse pasture runs between the house and the barn, and Franklin the pig has pushed his way under the fence and is rooting around among Maxx the gelding, a sweet old quarter horse, and his girls Star, Cajun, and Beauty. The horses are nonchalant, for they know the little piglet, currently the size of a large basset hound. Still, there's an obvious safety issue, and I wonder what new modification we'll have to make to dissuade the determined pig.

Lorraine is close behind the disgruntled animal. "Out right now!" she commands, and an annoyed Franklin scoots out, grumbling loudly. He passes just outside the kitchen, where Priscilla the potbelly snuffles the ground for remnants of breakfast rinsed from the horse, goat, and pig dishes. Franklin moves toward her, they conspire for a second, then in unison head quickly to the barn entrance. Paulie the rooster is there, too, enjoying the sun. Knowing that Franklin is misbehaving, he rushes to Lorraine's aid, flapping along behind Franklin and pecking at his behind.

"This really is my life," I say to myself, smiling.

Most of the animals are outside grazing; the barn is cool and peaceful. Rambo the sheep lies in the middle of the aisle; though resting, he's always watchful, and is the first to let us know when something is amiss. Hannah, his girlfriend, lies next to him, her woolly body pressed next to his. Hannah arrived six months ago. Found thin and filthy in a Queens cemetery, Hannah was terrified of human beings, so we thought it wise to keep her in the barn. First of all, she'd have more exposure to people than if she lived with the sheep herd at the far end of the property. Secondly, she'd be able to take her cues from Rambo, the greatest sheep who's ever lived.

Take her cues she did. Indeed, Hannah has become a loving and irrepressible spirit. Yet she's also fallen in love with Rambo, and on days when he needs some space and literally hides from her—behind the rabbit barn, behind the tool room, in the pig stall, or elsewhere—Hannah runs the farm frantically, calling out to her beloved who stands like a statue in his undiscovered hideaway. Once discovered, he shakes his head in frustration and stomps away.

I open the gate into the kitchen. "Boss alert! Everybody out of the hot tub, NOW!" Walt commands. April, covered from head to toe in muck, shakes her head, and someone else jokes good-naturedly about boiling Walt alive in the fictitious tub.

The large room that we call the kitchen is actually a multipurpose feed room, supply closet, and kitchen. Well, at least it's where staff and volunteers eat. There's no stove. Just a fridge for people food, and a larger one for produce—apples, carrots, plums, dark greens, pears, bananas—fed to the animals. These plus dozens of bags and bins of feed in all varieties. What makes it our lunch room is a massive table, donated by my beau, David, around which we congregate to down our lunches while Walt holds court . . . or at least, we indulge him in this way.

A five-foot wall divides the barn from the kitchen. We enter and exit via a wide gate that has kept the pigs at bay for a record length of time. Determined to claim the feed room as their own, the pigs have foiled several of Alex's initial attempts to keep them out. This time, his ingenious spring-loaded lock placed on top of the gate beyond their reach is working. Trust me, though, when something stands between a thousand-pound pig and a food source, that something is not long for the world. Alex knows this, and I'm sure he has Back-Up Plan #5 in the ready.

In warm weather, we wash feed dishes just outside the back entrance. A hot spray, a quick scrub and rinse, and dozens of rubber tubs are neatly stacked to await the next meal. The area is a favorite spot for the free-roaming pigs, chickens, and ducks. First of all, remnants of breakfast are there for the taking among dishes not yet washed. Secondly, the back door is also a gate, and coddled critters can look right through its woven wire and do what they do a little too well: beg to come in the kitchen.

Today, Petri the duck is just outside, her little beak sticking through one of the openings in the wire. Her quack is much more like a high-pitched, coarse squeak as she calls to April,

Lorraine, Walt, and Alex to let her in.

For the moment, they ignore her. "Do you know she won't take a treat until I pat her?" Walt asks. "Too bad the pigs aren't like that."

Paulie the rooster is outside, too, miffed that he's not at someone's feet, his normal noontime spot, demanding a bite of whatever we're having. Julia, a new hen, was rescued from the streets of Kingsbridge, a section of the Bronx, and having lived with humans for quite some time, chooses our company over that of other birds—especially, it seems, at lunchtime, when everyone gathers around the huge table to recharge and tolerate Walt's bad jokes. Today, it's her turn at the table.

"Baa-GAHHH!" screams Paulie. His neck feathers puff out in indignation.

"Oh, I feel sorry for Paulie," April says. "This is not right." She pushes her fit, tiny body away from the table, picks up Julia, and switches the two birds. "I'm sorry, girl, but Paulie's been having lunch with us for three years." A very grateful red rooster trots over to Lorraine's feet, stretches his neck as long as he can make it, and waits for a bite of her peanut butter and jelly sandwich. He downs it immediately, takes another bite, then struts over to Alex.

"Paulie, I didn't bring you anything today . . . I'm sorry!" Alex apologizes earnestly. Evidently Paulie doesn't believe him and pecks his calf several times. "I'm not kidding!" he continues. "I'll bring you a treat tomorrow. I promise." Paulie moves to the end of the table where Michelle, a new volunteer, sits.

"Well, what do *you* have for me?" Paulie eyeballs her.

- 16 -

Paulie the Yogi

We don't believe in either carting around or confining animals for show, and are therefore opposed to circuses, petting zoos, and other businesses that use animals for human entertainment. Common sense tells us that an animal, if given a choice, would choose to live in a natural environment like the one in which his species evolved and has lived for centuries. No animal wants to live in a small enclosure lacking any stimulation whatsoever, and no animal would choose to travel in tiny cages closed into darkened vehicles only to get out day after tedious day to perform the same tricks she's been coerced into learning. Surely most of us would agree that stress and boredom are not the sole purview of human beings, and that boredom, confinement, and suffering feel much the same no matter whether one is human or lion, alligator or bird. We should also consider what happens to circus and petting-zoo animals when they're old, injured, or hard to manage. Is there a Shangri-la out there to which these animals retire after years of stress and mind-numbing boredom? A place where animals subjected to the drudgery of such a life can enjoy spacious fields, cozy shelters, and kindness from caretakers? The answer is no. After a lifetime of enslavement, the vast majority of these animals are killed so that those who have used them for profit can squeeze the last pennies out of their hides, their meat, their bones.

So while we are constantly asked to bring animals to events, we politely decline. We tried it a couple times with our most social animals, who didn't enjoy being away from their friends, their home, their routine. So the animals now stay put, and people come to them.

Except for Paulie. The bird loves car rides. I knew this when several times in the space of a couple weeks he would charge out of the barn every time he heard my car. That he identified the sound and associated it with me is phenomenal enough, but then, once by the driver door, that he craned his neck as high as he could, flapped his wings and hopped up and down—well, yes, I laughed, shook my head, picked up my pal, and went for a ride.

Saying "Paulie likes car rides" is like saying "Murphy likes food." Murphy is obsessed with food. He sneaks it, steals it, devises ingenious ways to get into cabinets, metal cans, locked garbage bins, and onto countertops and tables. I once caught him barking incessantly at Policeman, hoping to scare the thousand-pound pig away from his breakfast (it didn't work), and we've all witnessed him stealthily exiting the barn with lunch bags, pizza boxes, and cakes.

So it is with Paulie, who will run out to my Subaru on his stiff old cockfighting legs, squawk his head off, flap his wings, and yell, "You'd better take me with you right now, lady, or you're in big trouble!"

Not always, but often, I do. Though his preferred seat is my lap, for safety reasons he usually rides shotgun, comfortably settled into a stack of folded towels. For the first few minutes he looks wide-eyed out the window, head twitching left, right, up, down as he takes in the big wide world passing by him. Soon he'll settle onto his cushion, though, and within minutes nod off to sleep. A good neck massage and a little classical radio make naptime arrive even more quickly.

Paulie has been to over twenty schools and several community events. He's a frequent guest on Susan Arbetter's *Roundtable* (WAMC Radio) and is a beginning yoga student at Manhattan's Jivamukti Yoga Center. Well, at least that's probably what he would tell you.

Sharon Gannon and David Life are world-renowned yogis and two of the most remarkable people I've ever known. Jivamukti, their own style of yoga created in the 1980s, is an extremely rigorous form based on the concept of compassion. Its mantra, "Loka samasta sukhino bhavantu," translates roughly to "May all beings, everywhere, be happy and free," and its creators, whose centers are located in New York, Toronto, London, and elsewhere, travel the globe to teach the yoga of kindness. I love them.

So Paulie and I were naturally thrilled to accept an invitation to celebrate at their Manhattan studio, speak about the work of Catskill Animal Sanctuary, distribute literature, and take a yoga class. Though he'd slept much of the way down during the two-hour drive, my chicken friend was nonetheless ready for a good stretch. Quickly I set up our information table. Jules Febre's yoga class was waiting for us. At Jules's request, I joined him in front of a class of thirty students, and gently placed Paulie on the floor to do whatever a rooster invited to yoga class wanted to do, which evidently was to practice yoga. In very deliberate fashion, Paulie walked toward a man in his thirties seated cross-legged on a purple yoga mat. Right onto the middle of the mat he moved, stretched his neck long and tall, and looked the student right in the eye.

"Aaah, he really likes you," I explained.

"I hope so," the good-natured student responded, smiling.

For the next few minutes, as the class discussed the work of CAS, Paulie sat contentedly at the front of the man's mat. Before long, he was sound asleep, his opaque eyelids closed over his wondrous copper eyes. What did these sophisticated urbanites think, I wondered, of the bird so content to be in their company? What did it mean that Paulie was every bit as happy right now sleeping on a purple mat in a building on Lafayette Street as he was scratching the dirt for bugs back at home? What did these people think when, after class, they held the gentle boy bird, a former fighting rooster, and felt him let go wholly into their arms, or when they watched him curiously exploring every inch of the yoga center one room at a time? What did they think when I told them to watch and listen for his expressions of joy as they massaged his neck or the soft exposed skin under his wings?

Of the thousands of people whom Paulie has met and acknowledged, how many have had epiphanies? Encouraged by his remarkable spirit, has anyone asked herself whether maybe, just maybe, her perceptions of animals are all wrong? And how many have asked the next question: Well, if this is who they are, do I really want to eat them?

In the appropriate environment, I can share with people what

Paulie and so many others have taught us, which is that perhaps it's not simply that Paulie is a remarkable bird, Babe a remarkable steer, Policeman a delightful pig. Perhaps it's also that we humans never know them in their infinite wonderfulness. For most of us, animals raised for food are nothing more than a convenient source of protein, and all most of us do is consider which parts of their bodies—their legs, their breasts, their rib cages, their wings—we want to eat. In hiding its hideous practices from us, agribusiness has successfully distanced us from the remarkable and unique and quirky cows and pigs and chickens who writhe as they are yanked helplessly into the air, scream mightily as, often fully conscious, their throats are ripped open so they bleed to death before becoming our hamburgers, bacon, or chicken nuggets.

For a brief moment, I contemplate a world in which animals are entitled to their lives simply by virtue of being born on our shared planet. What a glorious world it would be if most of us believed that it was not our right to use animals for any purpose or whim that humans chose, but instead, as Matthew Scully argues in *Dominion: The Power of Man, the Suffering of Animals, and the Call to Mercy*, that in giving us "dominion over the animals," God intended us to be their caretakers, their guardians, not their abusers, torturers, enslavers. Not their murderers. But it is not the time or place to make this speech. Another class is about to begin, and we've got rush-hour traffic on the West Side Highway with which to contend. So to the students delighted by the little red bird, I say simply, "Animals are so much more than most of us ever know," and leave it at that.

After several trips out to load up the car, I come back at last for Paulie. I pick him up, and with both hands hold him at eye-level right in front of my face, then kiss his beak. He stretches to move in closer, so with one arm, I cup him next to my face. Gently, so very gently, Paulie rests his head on my cheek. A small, taut young woman lifts her hand to her mouth. "Oh, wow," she says. "Look how much he loves you."

"Yes," I say. "Yes, he does."

Loka samasta sukhino bhavantu: may all beings, everywhere, be happy and free.

- 17 -

Road Rage

Friday afternoon, 4:00 PM. Not a good time to be leaving anywhere, especially not Manhattan. But there we were in my friend's pickup, inching our way up the West Side Highway. Paulie had enjoyed his big day at Jivamukti, but, like a child after a day at the beach, needed a nap. He pecked a few seeds and pieces of fruit from his bowl and dipped his head several times into his water dish, but within several minutes scrunched down into his towel and was sound asleep.

The West Side Highway on a Friday afternoon is not a fun place to be. It seems that of the more than eight million residents of the city, seven million leave every Friday afternoon: Yep, we love our city, but we can't wait to get the heck out . . . every single weekend. They were certainly leaving in force this springtime Friday afternoon, and I was suddenly regretting the decision to drive a manual Toyota pickup: first gear, second gear, neutral as we waited for the traffic ahead of us to move six whole feet, first, second, first . . . third gear, first gear, second. Somewhere in the Bronx the traffic lightened and we glided along for a bit at twenty-five or thirty miles per hour. I looked over at Paulie and smiled: this bird sleeps so soundly that I half expected a snore.

A black Volvo from the right lane jerked in front of us. No signal, just inches between our bumpers. Whoa! I slammed on the brakes and swore at this cliché—the mad driver who risks his safety and those of others to move a car's length ahead. Thrown onto the floor and out of his reverie, Paulie stood and shook his head.

"What in the heck was that?" I saw him thinking.

"Paulie! I'm so sorry, bird." Though he was clearly just angry and not injured, I knew that my continued chat would soothe him. "I am such a bad mama, Paulie! Can you believe what I did to you? I'm so sorry . . . are you okay?"

I needn't have said any of this, because Paulie was about to tell me exactly how he felt. He jumped onto the passenger seat, dropped his wing and started his "angry dance" toward me. And then it happened:

"SQUAA—AAAAWWK!" Louder than I knew he could screech, Paulie was yelling at me. "SQUAA—AAAAAWWKKK!!" He stared up at me and screamed at the top of his two-ounce lungs, and I swear, dared me to yell back. I actually think he was so angry that he was probably swearing. Yes, that's it . . . Paulie was using the f-word.

For a moment, my mind didn't know what to do with this. I was, after all, being chastised by a chicken.

"Keep it together, Kathy," I thought to myself. "You're surrounded by the world's rudest and most impatient drivers. You're in someone else's truck and Paulie is pissed." Somehow, suppressing the guffaws that were howling for release, I reached over to console my angry friend with a neck rub. For the first and only time ever, Paulie bit me, and then squeezed as close to the passenger door as he could to fall asleep.

§

May all beings, everywhere, be happy and free. Especially if they're roosters who have just been tossed to the floor of a pickup during a well-earned nap.

- 18 -

Add Water, Stir With Love

It's nearly dinnertime, and every single head in the barn is turned in anticipation toward the feed room: that secret sanctum from which all things delicious emerge. The donkeys stretch their necks out over their stall wall and stare, unblinking, at the gate that theoretically keeps out the four-legged. In their huge stall, thirteen goats stand at attention, their bodies angled toward the kitchen. They're listening, of course, for the opening of the gate, and then for footsteps moving in their direction.

Every animal in the big barn, in fact—eight horses, thirteen goats, thirteen pigs, two sheep, twenty-six chickens, nine rabbits, four ducks, and four donkeys—is rapt, listening to the sounds of lids slapping, grain plunking in buckets, knives clicking on the stainless steel prep table as the evening's treats—cantaloupe, kale, tomatoes, grapes, and bananas—are sliced. I half expect Jane the volunteer to emerge from the kitchen with a pad and pen and ask, "Would you like to hear today's specials?"

Hannah and Rambo stand side by side, heads touching, peering through the gate into the kitchen. Just behind them, Petri and Darwin, our two free-range ducks, wait patiently in the aisle. Gracie, an old gray mare, paces in circles in her stall, popping her head out and staring at the kitchen every few minutes. And then there's Franklin: no movement from him, just a nonstop "mmmph, mmmph," as he waits at his door, his pink snout pushed through one of the holes in the metal mesh gate.

At CAS, mealtime is magic. Walt designs the animals' meals to ensure optimum nutrition for every single individual, and a giant dry-erase board lists all ingredients for breakfast and dinner for 175 animals. It has more recipes than many cookbooks, and reflects a philosophy that might read something like:

§

FOR EVERY MEAL FOR THE REST OF HIS LIFE,
GIVE EACH ANIMAL THE VERY BEST FEED POSSIBLE,
NOT ONLY FOR HIS OR HER SPECIES,
BUT ALSO FOR HIS OR HER AGE, BODY WEIGHT,
PHYSICAL CONDITION, AND ACTIVITY LEVEL.
ALSO PROVIDE HEALTHY, LOW-FAT TREATS.
REMEMBER TO SERVE WITH LOVE.

§

The board is updated weekly as Walt tweaks diets when an animal is a tad over or underweight, a little stiff or creaky, dehydrated from drinking too little water, or even when his stool is slightly too firm or too loose. Heavy-duty plastic bins hold our staples, and dozens of other buckets, bins, bags, and boxes contain everything from rice bran to chopped alfalfa to glucosamine to Tahitian *noni* juice to prunes and bananas, all of which are used as necessary. Lorraine and April instruct volunteers to chop fruits and vegetables into bite-sized pieces for the pigs, and then like proctors hovering over students at test time, monitor the distribution to make sure that each pig gets an equal portion.

As I enter the feed room on this wet Saturday, Lorraine has nearly finished setting up for dinner. With the nutritional part of dinner measured, chopped, and poured for 175 animals, the last step is to add medication for those who need it—anti-inflammatories for Policeman the pig's arthritis, bute for a pony named Luna who's a bit off on her left front.

A squeak sounds from the aisle.

"Okay, sweetheart, I'm coming," she calls to Petri. Dinner is behind schedule and the little duck evidently knows this. Lorraine exits with the duck dishes and Petri follows at her feet, her two-inch legs a whir of frantic motion.

Franklin is getting louder. Next he'll rattle his door. And if we're especially slow, he'll begin to scream—a shrill, high-pitched scream that's the peculiar, ear-splitting forte of pigs.

April buzzes past. "Gary," she asks her devoted husband, who volunteers at CAS on his days off from programming at IBM, "Want to feed Franklin?" She hands off the rubber dish to an unwitting husband.

"Sure," he says.

"Here we go," I think. I'm a jerk for not offering to do it myself; instead, I decide to let Dennis the Menace share the wealth.

"Okay, everybody, I'm heading out," April calls from the end of the barn. April's job is to feed the outside animals, and, with bales of hay and buckets of grain stacked high in the truck, she's ready to go.

"Hon, is that enough hay for Claude?" she asks as she exits. Eric, a new volunteer, has prepared a cozy bed for Claude the pig, but looks uncertainly at April. "You might give him a couple more flakes, and shake them out really well. It's going to be damp tonight."

Standing outside Franklin's gate halfway down the aisle, Gary is resolute. "Hey, guy," he says to a screaming Franklin. "Sit." And Gary actually stands there, waiting for Franklin to obey. Apparently he has never had a pet pig. Franklin screams louder, jumps up on his hind legs and places his front hooves through the mesh gate.

"Give me my dinner, man!" he screams so loudly that I cover my ears.

Gary does not open Franklin's door, lest the 250-pound pig careen out at him. Instead, he hoists himself over Franklin's wall. Bad move. In today's game of pretend, Franklin is not Dennis the Menace. No . . . he's left such childish games behind. For Franklin, it's the last ten seconds of the Superbowl; the score is tied: Humans 14; Porcines 14. Gary has the ball. It's up to Franklin to make the tackle. He rushes in! The outcome of the Super Bowl is in his hooves! With a mighty leap he hurdles skyward, driving Joe Montana into the hay—I mean Astroturf—inches before the goal line! The crowd roars!

Gary climbs out. His limbs are intact, but he's white as a sheet. Franklin is eyeball-deep in his dinner dish.

§

Back in the kitchen, Paulie the rooster assists Lorraine by sampling an item—a pellet, a piece of kale, a grape—from each of the dishes arranged intricately on the floor. Volunteers Jane and Chris wait for instructions. Lorraine points to five stainless-steel dishes in a row. "Hampton, Zoey, Priscilla, Rosebud, Charlie," she explains, and Jane stacks them into the appropriate order to deliver them to the eager potbellies. Charlie is getting round again, so these days dinner consists mostly of fruits and vegetables. He doesn't complain. "Katie, Big Ted, Buddy," she instructs Chris, and Chris exits with each horse's meal. Lorraine herself comes out with two horse dishes, one for ancient Dino, one for Bobo, and the ritual proceeds in this way until all seventy-eight barn animals are fed.

For a few peaceful minutes, the barn is quiet. There is only chewing—the slapping of pig jowls, the deep crunch and grind of horses chewing—and the soft rustling of hay as goats, donkeys, and rabbits munch or settle in for the evening. It would be difficult for me to feed now, I realize with some sadness. It would take me twice as long as it takes Lorraine, April, Walt, and Alex. They've memorized those parts of the feed board that don't change, the locations of staples and supplements, the entire routine. As we've grown, I've pulled away from much of the daily routine, focusing instead on our legal cases against abusers, on our educational programming and work in schools, on overseeing the construction of new barns and shelters, and on raising funds to support our work. I miss the physical work long handed over to an extraordinary staff and their cadre of volunteers. I miss the stripping of stalls, the hoisting of shovel after shovel of wet bedding into the manure spreader to be replaced by fresh pine shavings or soft bedding hay. I miss stepping over chickens who are not afraid of the pitchfork or of me as the metal tongs scoop up the night's bedding. I miss the ducks quacking all the way to their pond as I scoot them out to clean their shelter, and I miss fluffing the hay to make warm beds for goats and sheep. In other words, I miss participating in the animals' lives as intimately as Walt, April, Lorraine, and Alex do. But here's the truth of the matter. First, I had to leave the barn in order for CAS to grow and thrive. Second, the animals are in the hands of four of the most loving people I have ever known. Nearly every week animals arrive hungry, sick, and often very frightened. These four people summon the best of themselves each day and do the work required to mend broken spirits, and once mended, to continue to love them mightily. They do it so very well, and the healing begins at seven every morning in the secret sanctum, that magical place from which all things delicious emerge.

- 19 -
Where the Blind Horse Sings

I hadn't used this saddle in years, but the curve of its seat on my arm, its weight, and its smell were so very familiar. "Want me to send your saddle?" my dad had asked on the phone one night.

It was the saddle I'd used as a teenager. Still, the curve of its seat on my arm, its weight, and its smell were instantly familiar. Now, though, rather than saddling "Hammer," my last jumper before I left for college, I was about to saddle a blind Appaloosa. My dad told me I was crazy. "The apple doesn't fall far from the tree," I replied.

Nothing about our work would change on this day. Nothing about our work ever changed. If Buddy seemed interested, we proceeded. If in any way I was pushing too fast, he let me know. I blabbed my way through placing a thick cotton saddle pad on his back. He didn't know what I was saying, of course, but from our first moments together Buddy had seemed both to understand and to trust my enthusiasm. So as I gushed about the fun we were going to have, I slowly placed the saddle on his back. Not only were Buddy's ears up; he was instantly ready to go. Well, my goodness. I tightened the girth under his belly and moved in front of him to massage his ears and utter some more encouragement. My friend didn't want encouragement. He wanted action. "Okay," I laughed as he nudged me out the door. "We're going!"

I had decided not to use a bridle. The bit—the metal piece that goes into the horse's mouth and attaches to the reins for control—seemed unnecessary. In the last week, Buddy had halted on a dime each time I commanded "Stop!" Along with two lead ropes clipped to the side of his halter, the verbal cues would be sufficient.

I led my friend down the drive and had to walk quickly to keep up with him. His ears were up, his head was high. There have been hundreds of moments in my life when I've wished animals had words to describe their feelings, and this was one of them. Buddy's behavior told me he was happy and excited; I sensed that he was also proud.

Between the goat paddock and the entrance to the woods was a cleared, fenceless area of about an acre or two. We walked the perimeter.

"Want to trot?" I asked. He did, and we trotted around once, and once more, cutting diagonally back to the center. "Stop!" I said. Buddy stopped.

The girth had loosened a bit. I tightened it a notch, and then it was time. "You ready, boy?" I asked. Buddy was. I gave him a good pat, placed my left foot in the stirrup, held the reins and top of the saddle, hoisted up, and swung my right leg over and into position. You can predict what happened next. Buddy immediately turned his head back, wanting a peppermint. I had a pocketful.

Good lord, this horse was too much.

On that first day, Buddy and I worked together for only fifteen minutes. We walked; we trotted; we circled; we jogged diagonally through the ring. We made tiny and giant figure eights. He tripped only once on an uneven piece of ground. "It's okay, Bud, just a little bump," I encouraged him, and he continued on. Halfway through I dismounted so we could play our little flinging game. Buddy flung me forward; he flung me again. I scratched his big old head that always seemed to itch.

I got back on for our cool down. Buddy had worked up a little sweat, probably more from nervous excitement than from fifteen minutes of work. Meanwhile, I was about to faint from joy. I lifted my left knee and loosened the girth. We circled the ring uneventfully—uneventfully except that Buddy was a blind horse, except that Buddy was so happy and proud and victorious that I was weeping my way through "Somewhere Over the Rainbow" before I caught myself and laughed.

§

How had such a spirit survived in a home that couldn't help him? From his first full day with us, Buddy seemed to be making up for lost time. Here was an adventurer; here was an animal with such joie de vivre that he often pushed me long after I was ready to stop. I rode him every other day. More would have been too much, I felt, for an animal of his age; less would have been a disservice to the indomitable spirit that was crying for release.

After a couple days in the ring, it was time to head to the

woods. Embraced by the sounds and smells of the natural world, this was where Buddy came alive.

"Down, down, down," I instructed, and Buddy took small, sure steps down the steep hill that leveled out at the stream. Before I needed to command "stop," he stopped. He heard the water dancing past us. Would he do it, I wondered? Would he cross the stream without me at his side? Buddy lifted his head high and to the left, as if willing his stone-blind eye to see.

"You can do it, boy," I suggested, and then waited, simply stroking his neck with my hand. If he was afraid, I'd be on the ground and guiding him through the creek in a split second. But if Buddy was afraid, he wasn't yielding to the fear. This extraordinary animal was summoning his courage and, for a few moments more, I would encourage him to do so.

"Whatever you want to do, Bud," I whispered. "Whatever you want to do."

Suddenly we were in the water. Step, step, step! Buddy lifted his feet high and placed them solidly and then we were on the far bank and navigating through piles of brush and fallen trees and then it was "up, up, up!" to climb ten or twelve steps up the steep rise and in fifteen seconds he had done all this, guided only by gentle tugs and encouraging words.

We followed the same trails as we had on our first woodland adventure. We walked the rolling terrain; we began up the slope. Two steps before the fallen tree, we stopped. I inched him forward, and though it took a few tries, Buddy navigated over the obstacle. No time to stop for a peppermint. Instead, a quick pat and we trotted onward. Every now and then, Buddy would stop suddenly, freeze in place, and listen. He'd wait for my reassurance, and then we'd advance. A few feet before the swampy area, I shortened the lead ropes connected to his halter to signal him to pay attention. He did. "Water, water," I explained, and he marched right through it, splashing both of us with water, algae, and muck. Indeed, our first ride through the woods was little different from one on an experienced sighted horse, except that Buddy couldn't see. Except that he was delirious with joy.

On each outing, Buddy and I would go farther than we had the day before. He demanded this. It took only a few rides for Buddy's occasional hesitation to dissolve. He was confident and eager, and each ride was magic. We trotted up hills, the leaves crunching beneath Buddy's hooves; we crossed the stream one day where it was narrow, and then the next where it was wide and deep. Often Buddy would lower his head for a drink and then just stand for a bit, enjoying the feel of the water. Aaah, it dawned on me one day, he can't see, so he's trying to feel. His blindness was the reason Buddy was so tactile. He couldn't see, so he enlarged his world not only through sound and smell, but also through touch.

One brilliantly sunny afternoon we discovered a clearing—it was actually someone's large backyard. Two hundred feet in front of the yard was a paved rural road that I knew led to open meadows of several hundred acres. Perhaps these fields belonged to our neighbor, Myron Langer. In any event, they were neither fenced nor posted, and Murphy and I had walked their wide open spaces many times. The terrain was smooth, solid, and as far as I knew, free of rodent holes that could easily snap a leg.

Hoping not to get caught, we headed quickly up the dirt driveway and hung a right onto the asphalt road. Buddy didn't miss a beat. In fact, he pulled hard against the pressure of the reins, for he wanted to run.

"Walk, Bud," I said as I held the reins short and tight and instructed Buddy to walk. Just another minute, I thought. We walked along the right edge while both our hearts settled a bit from the first fairly rugged portion of our ride. Each side of the road was largely wooded, punctuated only by an occasional house, until we reached the clearing.

Nothing but gently rolling hills were before us then. Far over the hills, perhaps close to a mile away from where we stood, lay the road that led back to CAS.

"Let's go, boy," I encouraged my eager friend. Grass the color of wheat brushed against Buddy's legs. "Let him collect himself," I thought, as we made a huge circle on the side of the first rise. But Buddy could barely contain his excitement. Neither could I mine.

And so I nudged him. Only once. Within four strides, Buddy moved from a walk to a trot to a canter. It was that simple: Buddy ran. My blind horse friend who just four weeks earlier had stood motionless in the driveway, afraid to move, was cantering, now galloping through a huge field of grass on a sunny autumn afternoon. Down a gentle slope we galloped, Buddy's front legs checking the length of his stride. He was confident and sure. When the hill turned toward the sky, I lengthened the reins and he ran hard, his breath keeping pace with each step, his front legs stretching long in front of him. On any other horse on such a glorious day, I would have noticed the wind on my face, or been aware of the kestrel swooping overhead as he hunted for field mice and rabbits. But not today. Today, the sun and the dry grass and the slope of the hill and the kestrel and my legs hugging this blind horse all coalesced to will him onward, triumphantly.

We ran freely for perhaps a half mile, and then Buddy began to tire. Slowly I pulled him to a walk. Rather than walk, however, Buddy stopped altogether. He gathered himself for a moment, and then, lifting his head high and proud, he belted out a triumphant "*nnneeeiiigggghhhhh*" that reverberated through the fields.

I am so alive! he sang.

Yes, absolutely. This time, when we were done, it was Buddy who sang.

-20 -

207 Happy Endings

I spot a parking space ahead of me in the crowded parking lot and squeeze between two vehicles. My stomach is in a knot; tears fight for release. I'm glad I'm wearing sunglasses.

To my left, a large, enclosed trailer hitched to a Ford F350 reads NEVES TAXIDERMY, INC. HUNTING CONSULTANTS. On my right, a black Dodge Ram sports a life-sized hand-painted tiger on its side, and a decal in the rear window reads "Safari International."

I rush to the entry, pay my ten dollar admission fee, start up the slope. In a pen on my right, a great white elk rests, his massive antlers reaching toward the sky, his legs folded under him. Just ahead of him in a dirt pen, a herd of dark, exotic sheep clusters together, far from the fence. They are frightened.

I move toward the sound of the auctioneer's drone: "Are you able to buy buy buy them for two and a half got forty-two and a half can you give me forty-five five five someone give me forty-five five got forty-five."

The auctioneer, fiftyish, is round and sports a crimson jacket; the spider veins in his alcoholic face match the color precisely. He is standing in the back of a pickup that inches forward from one pen of terrified animals to another as animal advocates bid against trophy hunters and taxidermists on this last sad day at Catskill Game Farm.

"One-seventy-eight's your buyer," the auctioneer barks.

I am forty-eight years old and I've never once had the thought I have now: these people are my enemies and this is war.

When the news broke that Catskill Game Farm was closing after seventy-three years, most of the public was devastated. In regional media, grandparents spoke of visiting the Game Farm as children every summer, and of returning with their children and their children's children. People described vivid memories of rhinos, parrots, alligators, camels, and of their delight in bottle-feeding baby sheep and goats.

"I felt like every animal in the world was represented there," one patron remembered. Businesses bemoaned the inevitable drop in tourism, and wondered what the options would be for long-term Game Farm employees. Headlines in New York City tabloids all decried the closing of CGF. When a blogger wrote that he and his family were "three generations of Catskill Game Farm lovers," and that "the day it closes its gates will be a very sad day indeed," he was expressing the sentiment of many who had known and loved the zoo.

Meanwhile, though, animal advocates were quietly celebrating. We had long believed that the Game Farm was anything but a happy haven for animals. Like most zoos, Catskill Game Farm kept its animals in small, sterile pens. Like most petting zoos, it lured the public every spring with scores of lambs and kids that visitors were able, for a price, to feed. But at the end of each visiting season, these animals were sent to auction, and thus to slaughter. Nor did the public know that Game Farm owners had been linked to canned-hunt operations and that over the years many of the most beloved and majestic creatures—animals who had never needed to fear humans—were sold to operations where they were shot at point-blank range.

Anyone with a trained eye could, on a quick walk through, see sickness and disease. Indeed, the United States Department of Agriculture, about as far as one can get from an animal-friendly organization, had cited the Game Farm many times for failure to provide such basics as veterinary care, clean drinking water, clean enclosures, pest control, and more. So while the public was mourning the closure of an institution, we were elated that what we viewed as suffering and exploitation would finally end. And yet there was an immediate concern: the fate of some two thousand animals who were to be auctioned off to the highest bidders on October 18, 2006.

My e-mail inbox filled with concerns: Is CAS doing anything? What can I do to help? What's going to happen to those poor animals? I suggested to a core group of savvy people that we meet, lest our efforts be diluted by working individually.

Ten of us sat on my back deck one August evening. We knew the likely fate of the animals: the farm animals, roughly half of the two thousand Game Farm animals, would go to auction and thus be slaughtered. Perhaps the Game Farm even had a relationship with Greenville, the slaughterhouse just north of us, and would ship the animals directly there. Some of the ponies, donkeys, and potbelly pigs had a chance of going to private homes, but just as great a chance of being bought by roadside petting zoos or traveling circuses. The exotic birds, we feared, would be bought by breeders, while hundreds of exotic "hoofstock" (Nilgai, Aoudad, rare deer, and others) would be snatched up by canned hunt operators and wind up as trophies on the walls of savage men and women.

We decided on two goals: first, to do all in our power to save as many lives as possible, and second, to raise public awareness of practices common to many zoos. Perhaps if the veil were lifted, the public would be less inclined to support places such as Catskill Game Farm.

I sensed that night what I'd sensed in the past several weeks: the hope that I would head the effort. I could not. CAS was embroiled in a tough and protracted legal battle against a horrible abuser whose attorney was defending her by attacking me as an "animal rights nut" who manipulated the entire case against his poor, innocent client who really did love her animals. I had to be careful not to fan his flame—both my name and the sanctuary's had to stay out of the newspapers. Jim VanAlstine and Kirsti Gholson, two experienced animal advocates, generously agreed to co-chair the group.

The first step was to try to convince the Game Farm owners to donate the critters to viable sanctuaries. In addition to saving their family legacy, they could take hefty tax deductions for the donated animals. They didn't budge; we hadn't expected them to. Jim wrote a series of press releases, and he and Kirsti handled the media with aplomb, deftly exposing the underbelly of the Game Farm business. A public protest was organized and held at the Game Farm entrance. Signs read CATSKILL SHAME FARM and 73 YEARS OF EXPLOITATION, NOT CONSERVATION. While the protest enraged the owners, they still dug in their heels. The auction would proceed as planned.

In the interim, our group, Advocates for Game Farm Animals (AGFA) had met, via the Web, with the Coalition for Game Farm Animals, formed by a D.C.–based woman whose goal was to find homes for as many animals as possible and to raise funds for their purchase. It was certainly not part of our original plan to purchase animals. Most of us abide by the party line that doing so only feeds into the entire system of animal exploitation. (As a purchaser, I simply put money back into the hands of the individual who will purchase more animals to exploit and slaughter.)

But in this case, we were willing to make an exception. First of all, the Game Farm was closing. The owners would purchase no more animals to exploit for profit and then, once used up, to sell for slaughter or hunting. Secondly, as Kirsti explained, "Public outcry was enormous. I got calls from around the country from people who were deeply moved and desperately wanted to help save animals." Thirdly, the drama of animal advocates bidding against zoos and trophy hunters certainly kept the auction in the news. If nothing else, it was an opportunity to educate the public about practices that zoos worldwide would rather keep under wraps.

Thousands of dollars were raised in the weeks prior to auction. Four women were our designated bidders.

§

The truck inches forward. "Step aside, clear the way," the auctioneer commands the crowd of some three hundred people. Slowly we part to let the truck through. Some of us follow the truck; others look for the placard that reads in six-inch letters, Next Auction Here.

"We got four piglets here," Spider Veins calls, "and we're going to start at a hundred dollars."

"Why don't you bid, Buck? That's a whole lotta pork chops."

I turn to see a fat man in fatigues and his friend in mirrored sunglasses and a baseball cap that reads, I'M AN ELKOHOLIC.

Mirrored Sunglasses says, "Nah . . . I got what I wanted."

The piglets, being piglets, are not frightened by the crowd. Rather, they poke quarter-sized snouts through the fence to greet each of us, and look directly into my eyes as pigs always do. I can no longer stem the tears.

"Two two two got two hundred anyone want 'em for two-fifty," the auctioneer hums. I lurch forward to Jim, who is monitoring the bidding. "Anyone got them?" I ask, and he knows that I mean, "Is someone who will love them bidding on them?"

"Yep, they're safe," he whispers.

And then around the bend to three calves, and my heart is again in my throat. They're male, several months old, with bloated bellies and ears crusted with blood where metal tags have been stapled. They live in a small pen, they have no grass, their mothers are long gone. They look at us with their immense doe eyes, all innocence. We crowd around them and the bidding begins. The smallest one walks up to me. I lean as far over the rail as I can and say, "Hey, little man. Hey there, sweet one." He takes my hand in his mouth. Only two people are bidding: a farmer standing beside me is eyeing the calves for slaughter, and someone else I cannot see. The hidden bidder is the final one.

"Let it be one of the good guys," I pray.

A rescue organization wins a group of llamas. The Coalition outbids a taxidermist for two porcupines, and a young couple with a pond places the final bid on a small flock of ducks. My rage subsides a bit.

My program indicates that a young miniature donkey and miniature horse pair are next. Our group's private notes read, THESE TWO ARE BONDED AND MUST BE KEPT TOGETHER. The bidding for the donkeys is over way too quickly, and I hear that the bidder numbers for the horse are different than they were for his pal. I rush up to our bidder, a woman named Barbara Cozzens. "Get that horse," I instruct. "I'll put in $1,000 of my own."

Four days later, our hauler Corinne Weber pulls down the driveway towing a dazzlingly white trailer with "Equine Emergency Transportation" painted on its side. The logo always makes me smile; Corinne and her daughter, Stephanie Fitzpatrick, have hauled at least as many pigs, goats, cows, and geese for us as they have horses. We've prepared two quarantine stalls at the far end of the barn. Alex has erected the portable quarantine gate so that none of our free-rangers can access the area; volunteers have bedded the stalls generously and filled water buckets and hayracks.

The pair of minis are the first off the trailer. They've lived their lives in a tiny pen in the Game Farm's nursery, where people purchase crackers to feed baby animals. They are both obese. We are surprised—shocked, really—by their fear and resistance. The horse, no more than thirty inches tall, rears straight into the air when we try to lead him into the barn. His pal, a donkey mare, does what all donkeys do when they don't want to move: she doesn't. Not a millimeter. Eventually, I get behind her (she, too, is a tiny thing), and lift her rear end into the air so that, little by little, she's forced to tiptoe to her stall.

Corinne heads to the quarantine paddock behind my house. She turns the rig around and backs up to the gate. We lower the trailer ramp, and there they are: three frightened calves. The good guys did get them, after all.

Why am I crying? Perhaps because this is a world that brutalizes cows, or because, in their youth and innocence, these three remind me of the innocence of all cows. Perhaps because the worst moment these three will ever know from now on is a late dinner every now and then, and that fact stands in such stark contrast to the conditions we force billions to endure. Perhaps because these animals (and 204 others bought by our group) are tangible symbols of the goodness, caring, and hope that sometimes seem to have disappeared from this world. Or perhaps I'm crying simply because three didn't die.

We named them Rudy, Amos, and Jesse. I kiss them every day.

-21 -

Saying Good-Bye

Five days after Paulie's death, Lorraine still has the corner of a peanut-butter-and-jelly sandwich in her bag. "I'm just not ready to let go of him," she explains, referring to her lunchtime ritual of sharing a piece of her sandwich with Paulie. He would only wait a moment; if Lorraine took too long to pull his treat from her bag, Paulie would cock his head sideways and stare upward at her, signaling her to hurry up. "He was like a dog . . . or a human, maybe . . . what an ability to communicate," she says wistfully. "I miss him. I miss how he groomed the pigs, I miss how he patrolled the barn—he liked a quiet barn, you know," she told me, explaining how Paulie had no tolerance for pig arguments and would rush right between the two fussing pigs, squawking and flapping his wings to tell them to stop fighting. Whether he objected to the noise (Lorraine's theory) or to the fighting (my theory), Paulie was a special friend to all of us. We will never forget him.

For those of us who love animals, it's a hard truth that their lives are rarely as long as we would like. We watch our beloved pets pass from youth to doddering old age in the blink of an eye. With the exception of horses and donkeys, who can live well into their thirties or even older, few farm animals live beyond their teens. Many animals are already old when they arrive at CAS, and so many others—the pigs and the broiler chickens, especially—are so biologically altered in the name of profit that their morbidly oversized bodies give out long before those of a naturally raised animal.

It's a sad fact of this work that we've grown accustomed to death. But no matter how many times we say good-bye, the process is never rote. We have loved each animal fully, and each death leaves its mark.

§

If ever there was one, Samson was a gentle giant. A Holstein steer, Samson was supposed to be turned into veal at a few months old. But he wasn't. We know little about his history prior to his seizure, along with sixteen other animals, from a local hoarder. We can only talk about his final year, the year with us, and what his life and his death taught us.

Samson stood a full 6'2" at the shoulder. His massive head surely weighed three hundred pounds. His body weight? Who knows: farm animal weights are estimated with cloth measuring tapes which are supposed to be accurate within twenty-five pounds. The largest tape we had for cows stopped at two thousand pounds, and it didn't even reach all the way around Samson's middle. We guessed he weighed 2,500 pounds or more. Though he could easily have killed any of us with one quick swipe of his massive head, the old boy was so happy to be with us, and wanted nothing more than carrots, kisses, and companionship, all of which he received in abundance. He also gave kisses, licking the faces of humans he loved as one sees mother cows affectionately licking their calves.

One summer morning, Samson lay in the grass when we pulled into work. Now there was nothing unusual about this: it was September, the grass was lush, and since we weren't luring him with carrots, there was nothing to compel him to come over to say good morning. Only Samson usually did. He had been Rambo's friend, seized from misery, and it was not just gratitude he conveyed; he certainly was the happiest and most affectionate cow—steer, to be precise—I could imagine. Probably for the first time in his life, he had acres of pasture to enjoy, the friendship of our small herd of cows, consistent food, and affection from humans. I believed he was both happy and grateful.

An hour after we arrived, Samson was lying in the exact same spot, placidly chewing his cud. I went out with a bucket of grain. Certainly it would help me confirm whether something was wrong.

"Big man, good morning!" I called, and placed the grain a few feet beyond his reach. Samson didn't attempt to stand up. Something was definitely wrong.

Later that day, the vet wasn't fazed by the fact that Samson had only moved eight feet all day long. He felt nothing in Samson's leg or hip to indicate anything more than a bruise or a slight tear. "Remember he's a huge guy, and an old one," he said. "It's probably just his instinct to rest. Let's give him until the morning."

Samson never walked again.

Perhaps, in retrospect, we waited too long to put him down. If we did (and I'm not convinced of this), it was only because he seemed as content as he could be out there lying in the grass. He tolerated effort after futile effort to help him stand: slings, hay bales wedged under different parts of his body . . . even tractor lifts. What a trooper he was when we rotated him five or six times a day (it took five strong people to accomplish this) to keep his blood circulating. And the gentle giant certainly didn't mind the attention lavished on him by an expanding group of well-wishers. Who knows whether in his last week of life Samson gave or received more kisses. I just remember a lot of kissing going on, a lot of carrots being consumed. The wise—and, yes, grateful—boy was making up for a lot of lost years.

And then one October morning Lorraine came running into the barn. The sky had just opened up, and Lorraine was saying, "Kathy, Samson's ready to go."

I ran out, and under the structure we had erected to keep Samson dry, I saw, for the first time since we'd discovered him down, an animal in distress. "Okay," I whispered, bending to take his big head with its cockeyed horns into my arms. "I'll call the vet."

He arrived an hour later and saw the hulking, immobile animal propped up by bales of hay and surrounded by bags of carrots, apples, and a dozen red-eyed humans. Whatever the vet thought, which was likely not what we'd have hoped, I'm grateful that he didn't voice it. Instead, he drew an overdose of tranquilizer and asked for a halter. "He'll fight this. Cattle don't like injections," he explained.

"No, he won't," I said with certainty. "He's ready."

Many who were on hand to welcome Samson when he first arrived at CAS were present to say good-bye. We knelt around him and embraced the great beast until every part of his body had a human hand on it. I sat at his head, rubbing his giant cheeks. "You've been such a good teacher, big boy. Such a good, good friend. Thank you for the lessons. Thank you for the love," I whispered. People sang, and I heard a hundred "I love yous" as Samson happily munched a few last carrots. He was utterly serene; utterly at peace.

And then two things happened that none of us will ever forget. First, when the vet approached him and knelt to inject the tranquilizer that would put Samson into a deep sleep before the second solution—the one that would stop his big old heart—was injected, Samson, whose head had been resting in my lap, turned it as far as he could to the left and back . . . so far that his wet nose nearly touched his shoulder. There are only two reasons he could have done this. In that moment, the jugular vein into which both solutions would be injected was clearly exposed. Either he somehow intuitively knew that this was how to help by exposing the jugular, or he eyed the stranger, the doctor, and was saying, "No, I'm here with my family. You don't belong here."

The huge syringe of barbiturate emptied into him, and then the second remarkable thing happened. Samson turned around, looked as deeply into my eyes as any living thing before or since, and licked my face. Then he licked it again, and again, and again. Over and over as he was dying, the cow who had finally known love was, I believe, saying thank you. He was, I believe, saying, "I love you, too," and he was, I believe, saying good-bye.

We buried our old friend with a bale of hay and a bucket of carrots. A few nights after the burial, Jesse and I stood at his grave and watched a brilliant orange setting sun cast long shadows over the valley. "You know," Jesse said of the beast named after Samson, the Israeli judge with superhuman strength, "these are strong shoulders to build a sanctuary on."

After his death, an artist painted a life-sized portrait of Samson. It hangs above the barn entrance, the words PEACE TO ALL WHO ENTER HERE engraved above it.

-22 -

Welcome to CAS

One hundred feet from the large perennial garden in the bend of the driveway, Bambi, Ricki, and Luna graze in the hill pasture. Their rumps turned toward each other, they flick their long tails out of habit, for on this breezy day, flies aren't lighting on vulnerable horse flesh.

I round the curve. Just over the hill that divides the parking lot from our largest chicken yard, Oreo, one of four miniature roosters, belts out his rendition of the opening of Beethoven's Fifth Symphony from his perch high up in the willow tree.

"That little fart's got a good set of lungs," I can hear my Dad saying.

Sumo, an iridescent rooster ten times Oreo's size, surveys the farm from the spot he's dug beneath the tree. He calls a scratchy response to Oreo, and is followed by Kinsey, a hundred yards away in his house near the cows. It's a myth that roosters only crow in the morning. Sixteen roosters call CAS home, and as long as they're out and about, they always have something to say.

I park in the shade and rest my back against the big willow at the edge of the goat pasture. Murphy, my yellow lab, flops down beside me, his back pressing against my leg. A potbelly pig whose tail never stops wagging saunters toward us.

"Mmmph," she greets us, and presses her cool wet snout into my thigh.

"I love you, pig," I say.

"Mmmph," she says back.

In the goat pasture, three goats balance on their hind legs and reach for willow-tree leaves, their favorite treat. Two others, Mufasa and Austin, play king-of-the-mountain on the boulder pile. Shorter, older, and far more portly than his pal, Mufasa doesn't have a chance. He knows this, but continues playing happily.

Suddenly, Hannah the sheep marches outside. Her head is high, her strides long and swift.

"Good morning, beautiful sheep!" I greet her. She acknowledges us with a slight turn of the head in our direction but doesn't slow down. She's looking for her Romeo, Rambo, who has evidently had enough of her neediness this morning and is standing motionless behind the rabbit house. Hannah found his first hiding place behind the tool room a few weeks ago, but for several days running, Rambo is undetected in his back-up spot. Visitors laugh when we introduce Rambo as our sexually harassed sheep. But in fact, he is. Rambo loves Hannah, for sure, but definitely needs his space—something that Hannah is reluctant to give him.

"Murphster!" a volunteer named Dee calls as we walk into the barn, but before she can greet the yellow dog, calls out "Wait!" to Alex, who's preparing to pull the tractor forward. "Eliza's under the tractor." She reaches underneath for the white duck with the broken beak. Eliza quacks a half-hearted protest and waddles away from Dee toward the pond.

"Everybody out?" Alex asks.

Julia the hen and Sammy the rooster are courting at the far end of the barn aisle. Rambo and the other free-rangers are all safely out of harm's way.

"All clear!" Dee calls.

Walt walks in holding a hen with a blue bandage on her foot.

"Hey, Dee?" he says.

"Yep?"

"Would you please ask Paula not to put straw in the bunny house? We use bedding hay for the rabbits, not straw."

"You bet," she says, and returns to her task. With twenty stalls and fifteen shelters to clean, there's not a lot of time to chat.

Inside the large room that does triple-duty as our feed room, lunchroom, and medical-storage room, Nutmeg the hen is now upside down on Walt's lap. Her eyes are closed, and she's as relaxed as if she's being massaged. He's treating an abscess on the bottom of her foot with Ichthamol, a tarry salve that draws out infection, and has just given her 0.4 ccs of Baytril for pain.

"She's a sweetie," Walt says.

I grab a bucket, fill it with bananas, apples, and carrots, and head out to greet our newest arrivals: four donkeys. Lorraine and Alex have traded places on the tractor, and this time, it's Lorraine who backs it into the barn, the manure spreader empty and ready for the next load.

"Hey, toots," says Lorraine, smiling at me. She's been with CAS from the beginning, and what a pleasure it's been to watch her evolve from someone terrified of horses and cows to an excellent caregiver skilled at diagnosis and treatment of a wide variety of ailments in all species. She's calm in a crisis, good at managing volunteers, and her fear of the large animals is long gone. "Where are you going?" she asks.

"To kiss our new friends," I explain, referring to the four donkeys who arrived the previous night. Among our favorite CAS duties, kissing the animals ranks high on the list.

"Oh, they'll be happy to see you," she says, smiling. This is one of the many wonderful traits of donkeys. In fact, someone should create a bumper sticker that reads FIND YOUR BLISS: GET A DONKEY, for they truly are the happiest of creatures. This is particularly true when humans are around. Indeed, wherever you are is where they want to be. They are underfoot, in your space, and usually nuzzling you or checking pockets for treats.

Murphy and I walk down the long lane that will take us to the newly built quarantine barn. On our way, however, we pass the pigs, rabbits, hens, elder horses, and cows, each species in its own spacious area designed with its needs in mind. The pigs, for instance, fare poorly in weather extremes, so their large barn has radiant heaters in it for extra warmth in winter months, and their pasture slopes down to and includes a small, shallow portion of the pond, so that they can cool themselves during July and August. The rabbit house was built directly under a large willow, as rabbits, too, are intolerant of heat. Their large yard is fenced by woven wire buried deep into the ground, so that these critters can burrow and dig to their hearts' content without our having to worry about escapees. Lovely, rolling hills comprise much of the farm, but our eight-acre elder horse pasture is flat and easy on old, arthritic joints.

Murphy and I stop to greet the rabbits. To my delight, Rosie the pig and Eliza the duck are bringing up the rear, and, more than likely, plan to join us on our walk. Rosie is a newcomer. A blue-eyed pot-belly, she is a 150-pound love package. Her sidekick, Eliza, is a fine duck mutt rescued from the New Jersey streets.

"Come on guys, let's go meet the donkeys!" I encourage them, for it's a bit of a hike for our portly pig pal.

A cluster of white hens rests under a willow tree in a second chicken yard. Immediately they start to talk, and I imagine what they're saying to each other. Do they wonder if I have some kale or bits of apple? While they're probably simply hoping for treats, I wonder if they're gossiping about our little hiking group. How often, after all, does one see a duck, a dog, a pig, and a person out for a stroll?

We stop at the pond, standing beneath one of thirty gracious willows that surround it. I heave not one, not two, not three, but four large sticks, one at a time, as far as I can throw them. Murphy does his best belly-flop ever and heads toward the first stick.

"Mmm-mmph," Rosie comments and looks as if she's considering joining him. Darwin, meanwhile, is utterly uninterested in the water, and merely stands there preening.

The mutt collects far more wood than a dog should be able to—he's done this for ten years and has mastered the art—and swims back toward shore, sticks poking at odd angles out of his mouth. "Good job, doggy! What a great get-the-stick boy you are!" I praise him. He drops the sticks, trots off a few steps, and throws himself onto his back, all four legs wiggling and spazzing as he does his "happy dance" in celebration of his dog-ness. Rosie the pig is fascinated and trots over get a closer look. An uncertain Murphy quickly sits up, and for a few seconds, black pig and yellow dog are nose to nose. Rosie, I think, may be Murphy's first pig friend.

We round the corner and greet the cows grazing in their large field. "Good morning, cows!" I call. "Hi, Babe! Hi, Molly! Caleb, what are you doing?" I ask. Generally one or more cows will walk over to say hello. But not this morning. It's only their third day in a brand new pasture, and grazing down the lush grass is serious business.

And then we see them, all ears and eagerness. The four donkeys—Twister, Cody, Mozart, and Hal—have spotted us coming their way, and are crushed into the near corner of the quarantine paddock. I hear the wind-up that's the prelude to a donkey's "hee-haw"—a grindy, scratchy inhaling as if they're trying to gather strength for what's to come.

"Boys!" I call. "We're so happy you're here!"

They step back just a few feet, and I press my way into their paddock to greet them. "Wait for Mama, dog," I instruct the mutt. Knowing this drill well, Murphy plops down in the shade a few feet from the gate.

I know it will only be a matter of days before any donkey greeter will be instantly enveloped by love, before the donkeys compete to be the one physically closest to the human visitor, before one has to be very mindful not to get a toe stuck under the weight of a hoof. For now, they don't know us, so they don't crowd me when I go in. But they do stand at eager attention, ears pressed forward in enthusiasm, and when I kneel and reach out to them, two of them nuzzle my hand.

It's been fifteen days since our last arrival, a little hen found on an exit ramp off the New York State Thruway, but I've long committed to memory the words I whisper to each new arrival the moment we meet.

"Hello, animals," I say to the foursome as I kneel in front of them. "We're so happy you're here. And you know what? You will never, ever be unhappy again. Only love, from now on, for the rest of your lives."

Now I'm not so delusional that I think the animals understand our words. But at CAS, our hearts are wide, wide open, and I hope they feel what we mean. Almost always, I think they do.

Epilogue: The Journey Continues

Big Ted arrived a few weeks ago. An ancient, massive draft horse with limited vision, he was to be euthanized unless we took him. His basic care—hay, senior feed, bedding for his stall, glucosamine for old joints—will cost close to $400 each month, so the decision to accept him wasn't exactly a practical one. Sometimes, though, we toss out the practical considerations. If anyone deserves a happy ending, it is Ted.

A jet-black Shire with a white blaze, Ted endured the typical draft horse's life, first pulling a carriage, then pulling plows in Amish country, then pulling wagons loaded with trees through the woods when he worked for years as a logger. Finally, he was sold to a third-rate riding camp that rented him out for trail rides—long trudges up and down a mountain—after which he collapsed, exhausted, to spend the night in a narrow chute barely wide enough to contain a two thousand-pound horse.

"He's afraid of people, and he's aggressive with other animals," the woman who had bought Ted to save his life told us. "And oh, yeah, he's a weaver." A weaving horse rocks rapidly from one front foot to the next, his head moving from side to side with each shift of weight. It's like pacing in place, and it indicates extreme anxiety or boredom, two things Ted had surely experienced in his long working life.

"We're going to have our hands full," I told the barn.

The trailer pulled up on a Thursday morning. I stepped up, greeted our new friend, and led him out to his new life. "Welcome to CAS, Ted. You're not going to know what to do with all this love!" I whispered over and over as we walked the farm to allow Ted time to relax. Rosie the pig grunted a hello as we passed by, and the goats pressed their noses hard against the fence. "Hi, new guy!" their posture said.

When Ted walked into his stall a few minutes later, he immediately turned to the corner and began to weave. Though I tried to calm him, he had shifted into another world. He was using the only coping mechanism he knew to manage his fear, and it was heartbreaking to watch. If only we could explain that his worrying days were finally over.

I hoisted myself onto the front wall of his stall and sat. Alex and April joined me. We wanted the old horse to hear kind voices and to feel our calm and happy energy. We talked about the big boy, his history, how we might help him.

And then something remarkable happened. Within minutes, Big Ted turned around, pulling with him not a plow or a wagon full of trees, but a shattered spirit. He stepped toward us tentatively, and with what seemed to me an incredible leap of faith, rested his giant head on my thigh. April, Alex, and I locked eyes.

Perhaps Ted sensed that he'd at last found home.

§

In our rescue work, responses such as Ted's are the greatest compliment we can receive, and we're noticing that they happen more than ever before. Coincidence? I don't think so. The hearts of the people who work here are wide open and colossal. They are patient hearts, too—hearts that allow each animal the time he needs to know in his bones that he'll never be hurt again. Animals feel this: some, like Ted, feel it right away.

And then there are the animal residents. After all, when frightened newcomers take their first steps onto CAS ground, they see more than smiling humans. Often they are greeted by Rambo and Hannah, who trot out to the driveway to see who has arrived; by potbelly pigs strolling through and wagging their tails. They're likely to see a volunteer walking down the aisle holding a duck or a chicken, kissing it as she moves it to safety. As our new friends round the far end of the barn for the first time, the goats rush in from the field to press their damp noses through the fence to say hello. CAS breathes love, and that is why broken spirits heal so quickly here.

§

Our mission is the same as it's always been: to save farm animals from abuse and abandonment and to raise awareness of their plight and its impact on all of us. If numbers are a measure, we've succeeded beyond our imagining. Six years after we took our first animals, we have saved over 1,100 lives. We have three thousand active members, seventy active volunteers, and forty more whom we can call in a pinch. School children from around the region visit to learn about farm animals and their plight, and when they can't come to us, I go to their classrooms. On weekends, visitors ask good questions, so we talk not only about how cows digest their food, for example, but also about what wonderful mothers they are, and about the deep friendships that occur within a herd. If the audience is not too young, we talk, too, about the conditions these docile, affectionate animals endure in feedlots and slaughterhouses—the misery of their lives, the horrors of their deaths. This is the stuff that caring people need to understand. Often while we talk, Babe the giant steer licks our faces with his sandpaper tongue. Many visitors leave with a vow to give up animal products. There are no better spokesmen for the cause than our animals, who, having nothing to fear, reach out in friendship to visiting humans.

While I am humbled by our success, I have always believed that there are plenty of good, caring people eager to sign on to something worthy of their time—a belief that's reconfirmed when staff and volunteers walk into the barn each day, and by the wholehearted commitment of Gretchen Primack, Jean Rhode, and Chris Seeholzer, who, as three-quarters of our board of directors, accomplish more than most boards of a dozen people.

There is something, though, that stops me in my tracks. After six years of working intimately with animals, here's what they've taught me: in ways that count, they're no different than we are.

Of course, not all animals are exceptional. As with humans, in most farm animal species there are leaders and followers, the outgoing and the meek, the bright and the dull, the impatient and the persevering, the placid and the high-strung, the subdued and the comical. Nor am I suggesting that animals have the same capacities to reason, empathize, or suffer pangs of conscience as we humans do. Far from suggesting that every animal is a Rambo, I'm simply stating that farm animals are much more like me than I'd have ever believed.

What does that mean, exactly? First, like each of us, every farm animal—every chicken, turkey, duck, cow, pig—has a distinct and unique personality. Everyone who knows Franklin the pig loves his impish sense of humor, and everyone who knows Rambo admires his courage, sincerity, and candor. Darwin is a loving but demanding duck, Charlie is a surly, impatient pig, and Imelda the chicken is obsessed with shoes. Every animal who passes through our grounds has dominant personality characteristics that have nothing to do with his or her species, and if a visitor were to randomly select five animals and ask the staff to describe their personalities, we could all do so, and we would say remarkably similar things.

The day I finished reading *The Pig Who Sang to the Moon: The Emotional World of Farm Animals*, I posed a question to the staff: could anyone name an emotion that's not exhibited by animals? None of us could! Elation, exuberance, joy, worry, gratitude, jealousy, sadness: we see them all here. I simply would not have believed that a formerly abused ram would reject the company of other sheep in order to watch over the entire farm and its inhabitants. I would not have believed that a rooster would so crave physical closeness that he'd demand to get in bed with me, or that as he was dying, a gentle old steer named Samson would lick my face over and over until he took his last breath. But this stuff happens all the time. Either we pretend it doesn't and go blithely about our work, or we pause, consider what it teaches us, and share what we have learned.

It is truly humanity's loss that we don't know animals on an equal footing. But it's far more than a mere "loss" for the animal who is grown and killed to be your steak, your chicken nuggets, or your leather coat or couch. Our distance and disconnection from them, our apathy regarding their treatment, have coalesced in support of the brutality of agriculture.

In this regard, the only distinction that truly matters between the Paulies, Rambos, and Franklins of the animal kingdom and the billions of others grown to feed us is that the Paulies—those who have found sanctuary—have simply drawn the long straws. That's all: they're the lucky ones. Not more unique. Not more special. Not more deserving. They are here as ambassadors for the rest. When we purchase meat and dairy products, our choices condemn Paulie, Rambo, and Franklin's brethren to lives that most good people would not wish upon rapists or child molesters. Why do we accept such lives for innocent, loving creatures who are powerless to stand up for themselves? Why do our desires, no matter how greedy, base, or cruel, always trump theirs?

In *The Way We Eat: Why Our Food Choices Matter*, Peter Singer and Jim Mason speculate that "otherwise good people" make "bad food choices" because they lack access to information that would encourage them to choose differently. And in *Dominion: The Power of Man, the Suffering of Animals, and the Call to Mercy*, Matthew Scully, a political conservative, writes:

> When a quarter million birds are stuffed into a single shed, unable even to flap their wings, when more than a million

> pigs inhabit a single farm, never once stepping into the light of day, when every year tens of billions go to their death without knowing the least measure of human kindness, it is time to question old assumptions, to ask what we are doing and what spirit drives us on.

In fact, many other wonderful books written by people with far more expertise than I discuss the cruelty heaped upon our food animals by agribusiness, and the implications of that cruelty for the rest of us. Some of my favorites are in the Bookshelf that follows this chapter. Too many of us avoid such literature. We don't want to be upset. We don't want to feel guilty when we sit down to our steak dinner, our grilled chicken. We don't want to change our diets—not even if they're killing us.

§

But if you have been moved by the delightful creatures of Catskill Animal Sanctuary, then I ask you to read one of these books. For you see, it is easy to be gripped by stories of individual suffering: poor Ted, poor Buddy, poor Rambo. Visitors ask with stunned indignation how on earth someone could starve his animals or set a barn on fire. What kind of monster would do that? How could you possibly move away and lock your pig in a closet? Good people who would never consciously harm an animal are enraged by the suffering inflicted upon the animals they meet, and are overjoyed time and time again at their happy endings. But the suffering of the individuals we have rescued pales beside that of the 104 million Franklins killed annually in the United States alone, or the nine billion Paulies killed each year. What of their stories? What of our role in their suffering? Is there room in our lives to consider them?

Naturally, the devastation wrought by agribusiness does not end with the animals. Growing animals for humans to eat is killing our planet as well. Quite simply, as scientists and other professionals around the world are screaming, unless we alter our corporate consciousness and personal lifestyles profoundly and immediately, Earth as we know it will not survive.

I'll never forget the first time I heard these words. They were spoken by an ex-cattle rancher named Howard Lyman at an animal-rights conference. Later, his book *Mad Cowboy: Plain Truth From the Cattle Rancher Who Won't Eat Meat* helped me connect the dots:

> We must indeed consider a complex web of interrelated problems: air pollution, water pollution, land contamination, soil erosion, wildlife loss, desertification (the turning of verdant land into a condition resembling natural desert), rain forest destruction, and global warming. Humankind's profligate consumption of animal products has made a significant contribution to all of these ills, and it stands as the leading cause of many of them. Certainly these problems wouldn't disappear overnight if the world suddenly became vegetarian, but no other lifestyle change could produce as positive an impact on these profound threats to our collective survival as the adoption of a plant-based diet.

What? Eating meat is harming the earth? How on earth could it be? I had to know! So I read Howard's books. I read Matthew Scully's powerful work. I read *Beyond Beef* by Jeremy Rifkin, *Diet for a New America* by John Robbins (heir to the Baskin-Robbins fortune), and a dozen more. Each writer discussed the same issues; each writer came to the same conclusions. Rain forests are destroyed at two-and-a-half acres per second, largely to make room for cattle or for the soybeans and grains used to feed them. Half a million pounds of cow manure—several hundred times more concentrated than raw domestic sewage—is produced on a typical cattle feedlot every single day; the largest feedlots produce as much waste as America's most populous cities. Pig, chicken, and cow waste from massive factory farms leeches into our soil and water. Overgrazing, overcropping, chemical-based agriculture and deforestation have decreased topsoil depth to less than six inches (from twenty-one inches) in much of our cropland. Each statistic is more alarming than the previous one.

Even the United Nations is weighing in with a somber warning. In a report released in November of 2006, senior UN Food and Agriculture Organization (FAO) official Henning Steinfeld wrote, "Livestock are one of the most significant contributors to today's most serious environmental problems." The report pointed out that cattle-rearing alone generates more global warming greenhouse gases than emissions from vehicles. It is a fact: polar ice caps are melting, our natural resources are being poisoned and depleted, and entire species of animals are being extinguished—in other words, our planet is dying—in part because of agribusiness.

There is a final arrow in the quiver for those still unshaken by what they've read. Quite simply, meat and dairy consumption is harmful to human health. Heart disease, heart attacks and strokes, cancers, degenerative diseases, and diabetes are all dramatically lower in vegetarians than in meat eaters. Even the conservative American Dietetic Association acknowledges that "appropriately planned vegetarian diets are healthful, nutritionally adequate, and provide health benefits in the prevention and treatment of certain diseases." For those interested in learning more, I heartily recommend *The China Study: Startling Implications for Diet, Weight Loss and Long-Term Health* by T. Colin Campbell.

In the midst of all this bleakness lies the happiest, simplest of solutions. One can choose to respect all animals. One can choose to tread gently on our fragile planet. One can choose a diet far healthier than one based around fatty, antibiotic- and hormone-laced meat and dairy foods. Eliminate animal products from your diet. We can align our lifestyles with our values, and in doing so, can make the world a little gentler for the creatures who share it with us.

I am honored to be among these animals. Ever a teacher, I am also ever a student, and the lesson I learn over and over, that we're all the same, haunts me. I do not pretend to know how or why Earth and all its magnificent creatures were created. But I do know this: the world is hungry for kindness. Surely every living thing is entitled to happiness; surely all hearts deserve to sing. If I can contribute to that end by giving up the use of animal products and loving the broken spirits that come my way, those seem to be very simple choices indeed.

May all beings, everywhere, be happy and free. Catskill Animal Sanctuary is my one small effort toward that end.

What's yours?

Bookshelf

The books below certainly opened my eyes; I hope they'll do the same for you. The titles will help narrow your selection, depending on whether you're most interested in the lives of animals, the implications of a meat-based diet for your own health, or the ways in which agribusiness is harming the planet. And remember to visit Catskill Animal Sanctuary—the animals have *at least* as much to teach us as any book does.

- Barnard, Neal. *Eat Right, Live Longer: Using the Natural Power of Foods to Age-Proof Your Body* (Crown, 1995).

- Campbell, T. Colin and Thomas M. Campbell II. *The China Study: The Most Comprehensive Study of Nutrition Ever Conducted and the Startling Implications for Diet, Weight Loss, and Long-Term Health* (BenBella Books, 2005).

- Coe, Sue, and Alexander Cockburn. *Dead Meat* (Four Walls

Eight Windows, 1996).

- Eisman, George. *The Most Noble Diet: Food Selection and Ethics* (Diet-Ethics, 1994).

- Eisnitz, Gail. *Slaughterhouse: The Shocking Story of Greed, Neglect, and Inhumane Treatment Inside the U.S. Meat Industry* (Prometheus Books, 1997).

- Fox, Michael W. *Eating with Conscience: The Bioethics of Food* (NewSage Press, 1997).

- Francione, Gary L. *Introduction to Animal Rights: Your Child or the Dog?* (Temple University Press, 2000).

- Goodall, Jane with Gary McAvoy and Gail Hudson. *Harvest for Hope: A Guide to Mindful Eating* (Warner Books, 2005).

- Lyman, Howard F. with Glen Merzer. *Mad Cowboy: Plain Truth from the Cattle Rancher Who Won't Eat Meat* (Simon & Schuster, 1998).

- Lyman, Howard with Glen Merzer and Joanna Samorow-Merzer. *No More Bull!: The Mad Cowboy Targets America's Worst Enemy: Our Diet* (Scribner, 2005).

- Marcus, Erik. Vegan: *The New Ethics of Eating* (McBooks Press, 2000).

- Marcus, Erik. *Meat Market: Animals, Ethics, and Money* (Brio Press, 2005).
- Mason, Jim, and Peter Singer. *Animal Factories* (Crown, 1980).
- Masson, Jeffrey Moussaieff. *The Pig Who Sang to the Moon: The Emotional World of Farm Animals* (Ballantine Books, 2003).
- Masson, Jeffrey Moussaieff with Susan McCarthy. *When Elephants Weep: The Emotional Lives of Animals* (Delta, 1996).
- Midgley, Mary. *Animals and Why They Matter* (University of Georgia Press, 1998).
- Regan, Tom. *Empty Cages: Facing the Challenge of Animal Rights* (Rowman & Littlefield Publishers, 2004).
- Rifkin, Jeremy. *Beyond Beef: The Rise and Fall of the Cattle Culture* (Plume, 1993).
- Robbins, John. *Diet for a New America: How Your Food Choices Affect Your Health, Happiness, and the Future of Life on Earth* (H. J. Kramer, 1998).
- Robbins, John. *May All Be Fed: Diet for a New World* (Harper Perennial, 1993).
- Robbins, John. *The Food Revolution: How Your Diet Can Help*

Save Your Life and the World (Conari Press, 2001).

- Schlosser, Eric. *Fast Food Nation: The Dark Side of the All-American Meal* (Houghton Mifflin, 2001).
- Scully, Matthew. *Dominion: The Power of Man, the Suffering of Animals, and the Call to Mercy* (St. Martin's Griffin, 2003).
- Singer, Peter (ed). *In Defense of Animals: The Second Wave* (Blackwell Publishing, 2005).
- Singer, Peter, and Jim Mason. *The Way We Eat: Why Our Food Choices Matter* (Rodale Books, 2006).
- Spiegel, Marjorie. *The Dreaded Comparison: Human and Animal Slavery* (Mirror Books, 1997).
- Stepaniak, Joanne. *Being Vegan: Living with Conscience, Conviction, and Compassion* (Lowell House, 2000).
- Stepaniak, Joanne. *The Vegan Sourcebook* (Lowell House, 2000).
- Wise, Steven M. *Rattling the Cage: Toward Legal Rights for Animals* (Perseus Publishing, 2000).
- Wise, Steven M. *Drawing the Line: Science and the Case for Animal Rights* (Perseus Publishing, 2002).

We would love for you to join us!

Most of the funding for our work comes from ordinary people who give modestly once or twice a year. If you'd like to participate in the mission of Catskill Animal Sanctuary, log on to our Web site, where you will find information about membership, volunteering, animal sponsorships, adoption, and more.

www.casanctuary.org

(845) 336-8447

316 Old Stage Road

Saugerties, New York 12477